T0321481

HARDWARE ANNEALING IN ANALOG VLSI NEUROCOMPUTING

THE KLUWER INTERNATIONAL SERIES
IN ENGINEERING AND COMPUTER SCIENCE

VLSI, COMPUTER ARCHITECTURE AND
DIGITAL SIGNAL PROCESSING
Consulting Editor
Jonathan Allen

HARDWARE ANNEALING IN ANALOG VLSI NEUROCOMPUTING

by

Bang W. Lee
University of Southern California

and

Bing J. Sheu
University of Southern California

KLUWER ACADEMIC PUBLISHERS
Boston/Dordrecht/London

Distributors for North America:
Kluwer Academic Publishers
101 Philip Drive
Assinippi Park
Norwell, Massachusetts 02061 USA

Distributors for all other countries:
Kluwer Academic Publishers Group
Distribution Centre
Post Office Box 322
3300 AH Dordrecht, THE NETHERLANDS

Library of Congress Cataloging-in-Publication Data

Lee, Bang W.
    Hardware annealing in analog VLSI neurocomputing / by Bang W. Lee
and Bing J. Sheu.
        p.   cm. — (The Kluwer international series in engineering and
computer science. VLSI, computer architecture, and digital signal
processing)
    Includes bibliographical references and index.
    ISBN 0-7923-9132-2 (alk. paper)
    1. Neural networks (Computer science)   2. Neural computers-
-Circuits.   3. Integrated circuits—Very large scale integration.
4. Simulated annealing (Mathematics)   I. Sheu, Bing Jay.
II. Title.   III. Series.
QA76.87.L44   1991
006.3—dc20   90-19256
                                                                          CIP

Printed on acid-free paper.

Printed in the United States of America

Table of Contents

List of Figures

List of Tables

Preface

Rapid advances in neural sciences and VLSI design technologies have provided an excellent means to boost the computational capability and efficiency of data and signal processing tasks by several orders of magnitude. With massively parallel processing capabilities, artificial neural networks can be used to solve many engineering and scientific problems. Due to the optimized data communication structure for artificial intelligence applications, a neurocomputer is considered as the most promising sixth-generation computing machine. Typical applications of artificial neural networks include associative memory, pattern classification, early vision processing, speech recognition, image data compression, and intelligent robot control.

VLSI neural circuits play an important role in exploring and exploiting the rich properties of artificial neural networks by using programmable synapses and gain-adjustable neurons. Basic building blocks of the analog VLSI neural networks consist of operational amplifiers as electronic neurons and synthesized resistors as electronic synapses. The synapse weight information can be stored in the dynamically refreshed capacitors for medium-term storage or in the floating-gate of an EEPROM cell for long-term storage. The feedback path in the amplifier can continuously change the output neuron operation from the unity-gain configuration to a high-gain configuration. The adjustability of the voltage gain in the output neurons allows the implementation of hardware annealing in analog VLSI neural chips to find optimal solutions very efficiently. Both supervised learning and unsupervised learning can be implemented by using the programmable neural chips.

The Hopfield networks have been found to be very useful for many real world applications. Due to the feedback characteristics of the Hopfield networks, the solution often gets stuck at a local minimum where the objective function has surrounding barriers. Simulated annealing is a popular method to search for the optimal solution. The solutions by simulated annealing is close to the global optimum within a polynomial bound for the computational time and are independent of the initial conditions. The theory and algorithms for hardware annealing for achieving the optimal solutions in parallel have been developed. In hardware annealing, the voltage gains of output neurons in analog VLSI neural chips are increased from an initial low value to a final high value in a continuous fashion. Hardware annealing can be applied to pure analog and mixed-signal neurocomputing systems. It provides a speed-up factor of more than 10,000 times over simulated annealing in general-purpose workstations.

Overview of the Book

This book contains seven chapters.

Chapter 1 introduces artificial neural networks and associated VLSI implementation methodology. A brief discussion on the neural network properties is presented. The important findings and breakthroughs in neural network paradigms and architectures are reviewed. Design methods for various analog and digital VLSI neural chips are highlighted.

Chapter 2 presents the circuit dynamics of Hopfield networks. The existence of local minima in the energy function of a Hopfield network is examined. VLSI design methods to eliminate local minima are discussed. Applications of these methods to neural-based analog-to-digital converters and traveling salesman problems are described.

Hardware annealing theory and experimental results are presented in Chapter 3. Simulated annealing for software computation is reviewed first. The starting voltage gain and final voltage gain for hardware annealing are derived. The theoretical proof for hardware annealing is provided through the use of Gerschgorin's theorem. Experimental results on neural-based analog-digital converters by using hardware annealing are presented.

Chapter 4 presents compact designs of electrically programmable synapses and gain-adjustable neurons. The synthesized synapse cell consists of a transconductance amplifier to function as a three-terminal transconductor. The third terminal is used to control the effective conductance of the synapse circuit. The gain-adjustable neurons are used to facilitate the execution of hardware annealing in VLSI neural systems. The medium-term and long-term memory storages are achieved through the DRAM-style and EEPROM-style charge storage methods, respectively.

System integration issues are discussed in Chapter 5. A system module with programmable neural chips and standard IC parts is presented. Detailed system implementations for neural-based A/D conversion and real-time digital image restoration are described.

Chapter 6 presents alternative analog and digital VLSI neural chip designs. The neural sensory chips from Professor Carver Mead's group are discussed. Various industrial chips together with university research results are presented.

Conclusions and directions for future work are presented in Chapter 7.

This book is designed to provide engineers and scientists with an introduction to the field of VLSI neurocomputing. It is intended for use at the graduate level, although seniors would typically have all of the required background knowledge. This book is written to support a semester course.

The material of the text was used in graduate-level course entitled VLSI Neurocomputing in the University of Southern California in 1989, 1990 and in National Chiao Tung University, Hsin-Chu, Taiwan in 1990.

Los Angeles, California   *Bang W. Lee and Bing J. Sheu*

Acknowledgement

We would like to thank our colleagues in the Electrical Engineering Department and Computer Science Department at University of Southern California. Professor Hans Kuehl, Chairman of Electrical Engineering-Electrophysics Department, and Professor Jerry Mendel, Chairman of Electrical Engineering-Systems Department, promote a very creative and unique environment for VLSI research at USC. Dean Leonard Silverman and Associate Dean Lloyd Griffiths of the School of Engineering provide tremendous supports for our efforts. Professor Michael Bass, a former Chairman of Electrical Engineering-Electrophysics Department, and Ms. Ramona Gordon, the Administration Assistant, helped to initiate the VLSI Computing research at USC. We are very thankful to Professor Irving Reed for his invaluable insights and suggestions in VLSI computing. Professor Michael Arbib, the Director of Center for Neural Engineering (CNE), helps to spur the neurocomputing research. Professor Rama Chellappa, the Director of Signal and Image Processing Institute (SIPI), contributes to our knowledge in digital image processing. Professor Wlodek Proskurowski provides stimulating insights in the mathematical formulas. Our work also benefits from the lectures on neural networks by Professors Bart Kosko and Keith Jenkins. Comments from Professor Chung-Yu Wu, the Director of Institute of Electronics at National Chiao Tung University in Hsin-Chu, Taiwan, were of great value. Professors Paul R. Gray and Donald O. Pederson of University of California at Berkeley provide us long-term encouragement in analog and digital VLSI study.

Dr. George Lewicki, a former Director of the MOSIS Service at the USC/Information Sciences Institute and currently a Vice President at Orbit Semiconductor Inc., provided excellent supports and valuable comments from the beginning of our study in VLSI neurocomputing. Graduate research assistants in our VLSI Image Processing Laboratory, especially Joongho Choi, Ji-Chien Lee, Han Yang, and Oscal Tzyh-Chiang Chen, Chia-Fen Chang, and Sudhir Gowda, helped to obtain some experimental and simulation results. Mr. Joongho Choi also provides the key assistance to prepare the book in the camera-ready format. The fellowship support to Dr. Bang W. Lee during his Ph.D. study at USC from Samsung Electronics Company, especially under the arrangement of Vice President Kwang-Ho Kim and R&D Center Head Kwang-Kyo Kim, has been invaluable. The love, understanding, and encouragement from our wives Hae-Kyung Lee and Shelley A. Sheu are greatly appreciated.

This research was supported by Grants from USC-Faculty Research and Innovation Fund #22-1502-9759, USC-Biomedical Research Support Program from National Institute of Health #2-S07-RR07012-23, Powell Foundation #22-1501-5955, and AT&T Bell Laboratories #53-4599-0898.

HARDWARE ANNEALING IN ANALOG VLSI NEUROCOMPUTING

Chapter 1

Introduction

Recent advances in neural sciences and microelectronic technologies have provided an excellent means to boost the computational capability and efficiency of complicated engineering tasks by several orders of magnitude [1]. Due to the optimized structure for artificial intelligence applications, a neural computer is considered as the most promising sixth-generation computing machine. The interdisciplinary nature of neural network studies spans many science and engineering fields including neuroscience, cognitive science, psychology, computer science, physics, mathematics, electrical engineering, and biomedical engineering. Many digital neural coprocessors, which usually interface to personal computers and engineering workstations, are commercialized for accelerating neuro-computation. Typical products include ANZA from Hecht-Nielson Co., SAIC from Sigma, Odyssey from Texas Instruments Inc., and Mark III/IV from TRW Inc. [2]. However, a general-purpose digital neural coprocessor is usually much slower than a special-purpose analog neural hardware which implements the neural network in an optimal fashion.

Implementation technologies for the special purpose neural machines can be classified into two categories. The first technology is the electronic implementation using analog, digital, or mixed-signal chips. Several significant achievements have been made including early results from AT&T Bell Laboratories and California Institute of Technology.

By using a state-of-the-art MOS fabrication process, several hundreds of neurons and thousands of synapses can be realized in a single IC chip. However, the present silicon integration scale is still very limited as compared with biological systems. The other implementation technology is to use optical and optoelectronic devices. Due to the inherent parallelism and spatial property of optics, the interconnection and synapse weighting problems in multi-dimensional signal processing can be more efficiently addressed for certain applications. The high sensitivity problem on the portability of the opto-neural computer will diminish by combining electronics with optics.

The secret of immense computational power in neural networks is discovered as the parallel processing done by neurons and synapses. While each neuron performs simple analog processing at a low speed, the rich connectivity among neurons through synapses provides powerful computational capabilities for the large quantity of data. The data are processed asynchronously in the time domain and spread globally into all network elements. In addition to the parallel processing nature, the neural network has a self-learning capability which is done by changing the weights of synapses between neurons. The self-learning capability makes neural networks useful in a situation where training data are sufficient and fault tolerance of a system is necessary. Moreover, relatively imprecise network elements can be compensated with the self-learning scheme. The immense computational power and self-learning capability give neural networks excellent prospects in image processing, speech processing, artificial sensor processing, machine vision understanding, inexact knowledge processing, natural language processing, stock-market, forecasting, linear/nonlinear programming, and scientific optimization.

Figure 1.1 shows a block diagram of an advanced neural computing system. Real world signals are converted into discrete form (mainly digital) at the interface block. The neural signal processing system handles the converted signals, and the outputs can be transferred to digital computers for further data manipulation. The interface block might function as a data converter and conduct some early signal processing to the same level that human retinas can. Due to the robustness of neural networks, outputs of the interface do not have to be completely precise which is in sharp contrast to the conventional interfacing devices. The signals inside a neural signal processing system are distributed throughout the whole network. Thus, a small amount of damage in the system does not produce noticeable degradation of the overall system performance. Through self-learning procedures, the interconnection weights can be modified so that the original system performance is retained.

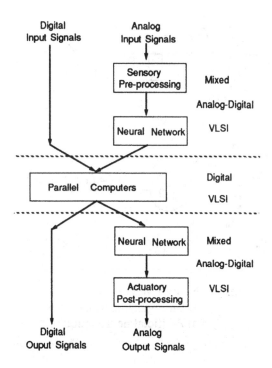

Fig. 1.1 Advanced neural computing system.

1.1 Overview of Neural Architectures

A biological neuron basically consists of a cell body, dendrites, and axons, as shown in Fig. 1.2. The cell body, which is called *soma*, performs complicated chemical processes, such as summation and firing with respect to a threshold level. The input signals for a cell body are transmitted through the dendrites, while the output signals are carried to other cells through axons. The electrical signal of an axon connects to a dendrite through a specialized contact, which is called the synapse. In general, the neuron performs a simple threshold function. When the potential inside the cell body is larger than the threshold value, the neuron fires. The normal firing rate is quite low, which is typically a few

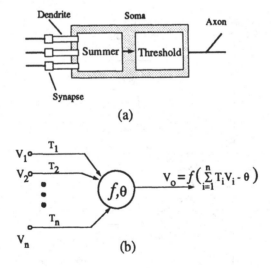

(a)

(b)

Fig. 1.2 Neuron Model.
(a) Biological neuron model.
(b) Artificial neuron model.

hundred occurrences per second. It is estimated that the human brain has approximately 10^{11} neurons and 10^{14} synapses. In the artificial neural network, the neuron and synapse are configured as processing element and connection strength, respectively. Various features of the artificial neural network are determined by the function of the neuron and interconnection patterns.

The artificial neural network can be characterized by the following properties:

> network architecture,
> retrieving process,
> learning rule, and
> training data.

The network architecture provides the most distinguished feature. The grouped neurons which are arranged into a disjointed structure are called *layer*. Figure 1.3 shows several architectures which include single-layer/multilayer and feedforward/feedback networks. The neuron transfer function and threshold voltage characterize the retrieving process of an artificial neural network. Specific mathematical functions including sigmoid, step, Gaussian, and Boltzmann functions are widely used to model the neuron transfer function. The nonlinear transfer function decides the information propagation properties at the neural retrieving process. The retrieving process can operate in either synchronous or asynchronous mode. In the synchronous mode, all neuron outputs are updated simultaneously. Conversely, the neuron updating process in the asynchronous mode is random and independent of the other neurons. Most artificial neural networks in software computation operate synchronously, while the biological neural network operates in the fully asynchronous mode. The training procedures are divided into *supervised learning* and *unsupervised learning*. In supervised learning, synapse weightings are tuned by the difference between the retrieving patterns and expected patterns. In unsupervised learning, the network classifies the input without References. The neural networks using unsupervised learning can detect the pattern regularities. The widely used learning rules include Hebb rule, Delta rule, competitive learning rule, Boltzmann learning rule, Hopfield energy minimizing rule, and their derivatives [3,4]. In general, the input signal for an artificial neural network can be discrete or continuous values.

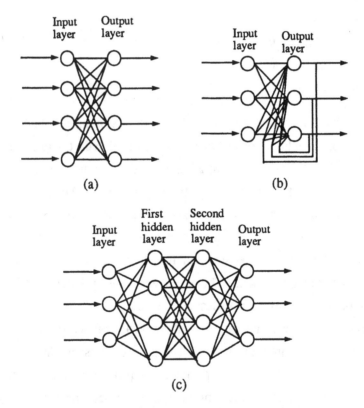

Fig. 1.3 Several neural networks.
  (a) Single-layer network.
  (b) Single-layer network with feedback.
  (c) Three-layer network.

After McCulloch and Pitts introduced the abstract neuron model for performing a simple task in 1943 [5], the neural network study began. The major neural networks and learning rules are listed in Table 1.1 and Table 1.2, respectively. F. Rosenblatt developed the *Perceptron*, which sparked a great amount of research interest in neurocomputing [6]. The Perceptron is a single layer feedforward network for pattern classifiers. Initially, the Perceptron demonstrated an optical pattern recognition when inputs of the system were connected to a grid of photocells. The input signals were then transferred to the neural layer with randomly weighted connections. The neural network performed successfully with application of the Hebb learning rule. The major limitation, pointed out by Minsky and Papert, is that the Perceptron cannot represent an exclusive-or function so that the Perceptron cannot classify complex categories. Multilayer Perceptrons were developed by Rosenblatt in the 1960's to overcome the limitation of the initial Perceptron.

In the late 1950's, the first neural hardware, called *adaptive linear element* (ADALINE), was developed by B. Widrow [7]. The neurons were realized with vacuum tube amplifiers, while the synapse weightings were manually adjusted with variable resistors. The ADALINE was improved to become MADALINE, which consists of ADALINE and a two-layer variant. These were adapted to a variety of applications, such as speech recognition, character recognition, weather prediction, adaptive control, and echo cancellation.

Another major category in neural networks is associative memory. J. Anderson proposed 'Brain-State-in-a-Box Model' with his linear associator and Hebb learning rule [8]. The network consists of a layer with feedback and one postprocessing output layer. Due to the positive feedback architecture and the learning rule, the output is the best-matched pattern from the stored memory for a given input.

Table 1.1 Major neural network models

Name	Years introduced	Primary Applications
Perceptron	1957	Typed-character recog., Simple pattern classifier
ADALINE/ MADALINE	1960	Adaptive processing; (equalizer, modem), Hand-written char. Recog.
Brain-State-in-a-Box (Linear Associator)	1977	Associative memory
Hopfield Network	1982	Image processing, Assoc. memory, Problem solver
Multi-layer w/ Back-propagation	1974-1985	Wide ranges: Speech synthesizer, Image Processing
Boltzmann Machine	1985	Pattern recognition for radar, sonar
Biderectional Associative Memory	1986	Associative Memory

Table 1.2 Basic learning rules

Model	Basic Equation	Comments
Hebb's Rule	$\Delta T_{ij} = \alpha\, V_i \bullet V_j$	Unsupervised, Hopfield network
Competitive Learning	$\Delta T_{ij} = \alpha\, V_i\, (V_j - T_{ij})$	Unsupervised, Associative Memory
Delta Rule	$\Delta T_{ij} = \alpha\, V_i \bullet E_j$	Supervised, Back propagation
Boltzmann Machine	Boltzmann probability function w/ simulated annealing	Supervised, Boltzmann Machine

In 1982, the presentation of J. Hopfield's paper to the National Academy of Science ignited the neural network study again [9]. The Hopfield network is basically a single layer with feedback. The condition for the synapse weighting is very restricted (being symmetric and having no self-feedback terms), while that for the neuron transfer function is very relaxed (only monotonically increasing and bounded). Using the energy of Lyapunov function, J. Hopfield proved that the network always moves toward a low energy level. Due to the simple architecture and clearly proved dynamics of the Hopfield network, many hardware implementations and real world applications have been accomplished. This network has been applied to associative memory and many engineering optimization problems.

The multilayer neural networks are vitalized by the back propagation learning rule. Before the learning rule was developed, the usefulness of the multilayer neural network had been well known, but the decision of synapse weightings was the main problem. The multilayer neural network can be used for various applications including data encoding/decoding, data compression, signal processing, noise filtering, pattern classification, and forecasting.

A Boltzmann machine has the similar network architecture as the Hopfield network, but differs in the stochastic update and learning properties. The stochastic update in retrieving and learning processes is based on the simulated annealing technique using the Boltzmann probability function. By decreasing the temperature of the probability function from a high value, the network always finds the global minimum in the energy surface.

The bidirectional associative memory (BAM) designed for optical computing is a generalized Hopfield model to heteroassociative network [10]. BAM has two fully connected central layers and input/output buffer layers. The synapses and neurons in the central two layers are bidirectional. For a given input, the BAM layers oscillate until a stable state is reached. The final stable output is the closest association stored in BAM.

Learning is the process of adapting the synapse weightings in response to external stimuli. The learning rules were developed with the network architectures. The first learning rule, named *Hebb learning rule*, which shows that the neural network can learn for a certain function, was presented in 1957 [11]. The rule requires that if an input and output are activated at the same time, the weighting between the input and output is increased. T. Kohonen developed *competitive learning*, where each neuron competes with others at a given input and the winner adapts to get more strength [12]. This type of learning, called *unsupervised learning*, does not need Reference data. On the other hand, desired outputs can be given in the *supervised learning* approach. The simple *delta rule* is applied to adjust the synapse weightings, using the error between the desired output and actual network output. Many derivatives from the simple delta rule are used for efficient learning results. The famous application of supervised learning is back-propagation for a multilayer network. The root-mean-square error at the output layer propagates backward through the network and is used to update the synapse weighting between layers. The process continues until the input layer is reached. The delta rule or its derivatives are used for determining the synapse weighting modifications. Due to an extremely long learning time, only one or two hidden layers are popularly used. The counter-propagation learning [13], which can be applied only to three-layer networks, combines the competitive learning and delta rule. Thus, the counter-propagation network selects the nearest stored pattern due to the competitive operation inside the hidden layer.

Another type of learning that falls between unsupervised learning and supervised learning is *reinforcement learning* [14]. In this learning, an external observer gives a response as to whether the network output is good or not. The learning rule of a Boltzmann machine [15] is based on the stochastic process, which constructs distributed representations of the Reference patterns with the simulated annealing technique. Due to the cooling process, the retrieving and learning time of the Boltzmann machine is extremely long.

1.2 VLSI Neural Network Design Methodology

The special purpose VLSI implementation of neural circuits and systems is extremely attractive in terms of size, power, and speed [16]. Various circuit design styles including pure analog, digital, and mixed-signal circuitries can be utilized to construct the VLSI neural chips. Since all network elements in the VLSI technology should be laid out on a two-dimensional floorplan, a significant limitation of the VLSI implementation is caused by a large number of interconnections and synapses. Thus, the existing programmable VLSI neural circuits are mainly restricted to compute only one-dimensional input information and to produce one-dimensional output results. When the input signals include two-dimensional information, a special mapping should be considered during the implementation.

The neural network algorithms for software computation cannot be directly used to realize analog VLSI neural networks. Basic elements of the analog VLSI artificial neural networks consist of amplifiers as neurons and resistors as synapses. The neuron transfer function for software computation is an exact mathematical function, while that for the VLSI implementation is an approximate function. Since the neuron transfer function is an important factor to decide network characteristics, some neural networks, such as the Boltzmann Machine and back-propagation, cannot be directly realized in VLSI design. In addition, perfect match and wide dynamic ratio of the synapse weightings are difficult to obtain in CMOS VLSI technologies.

One important factor in analog VLSI neural network design is to increase the amplifier gain. Figure 1.4 shows the simple circuit realization of a neural model using resistors. The synapse weighting T_i is realized with a conventional resistor R_i. For a negative synapse weighting, a resistor and an inverted neuron output are used. Several analog VLSI chips [17,18] have been reported using this direct resistor realization. The amplifier output voltage is determined by

$$Y = f\left(\frac{\sum\limits_{i=1}^{n} X_i/R_i}{1/R_{eq} + \sum\limits_{i=1}^{n} 1/R_i}\right). \tag{1.1}$$

Here, $f(.)$ is the transfer function of a neuron and R_{eq} is the input impedance of the neuron. On the other hand, the neuron output in software computation is determined by

$$Y = f\left(\sum\limits_{i=1}^{n} X_i/R_i\right). \tag{1.2}$$

In the direct resistor realization method, a large amplifier gain is needed

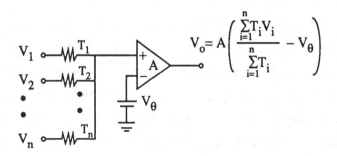

Fig. 1.4   A VLSI neuron with direct resistor implementation.

due to the effect of the denominator in (1.1). The relative synapse values are more important than the absolute values in the VLSI implementation. In VLSI technologies, relative tolerance in the device parameters is much easier to achieve than absolute tolerance.

The chip area and number of package pins can also be major limitations. The required chip area will be gradually decreased with advances in the fabrication technologies and with novel circuit design techniques. On the other hand, the number of package pins which is determined by the applications still remains the fundamental limitation. Each physical package pin can be shared by several functional outputs through time multiplexing. In this case, the analog VLSI neural chip might operate in the synchronous mode. Notice that original neural networks were developed to operate asynchronously.

For many applications, synchronous operation differs significantly from asynchronous operation. One good example is the use of a Hopfield neural network to solve a 10-city traveling salesman problem. Two versions of the forward Euler integration methods were examined: the fixed time-step version and the adaptive time-step version with Richardson extrapolation [19]. The former version corresponds to synchronous operation, while the latter version corresponds to asynchronous operation in VLSI neural circuits. For simplicity, the capacitances in the Hopfield network [20] are set to be 1 and the neuron inputs are initially reset. Figure 1.5 shows the energy function change as a function of time. The CPU time for the fixed time-step being 1.0×10^{-4} is about 5 hours, while that for the adaptive time-step with 1.0×10^{-3} error tolerance is only 0.5 hour. Here, the software simulation is done in a SUN 3/60 workstation. With the fixed time-steps of 1.0×10^{-3} and 1.0×10^{-4}, the energy function is no longer continuously decreasing due to numerical error from the synchronous operation.

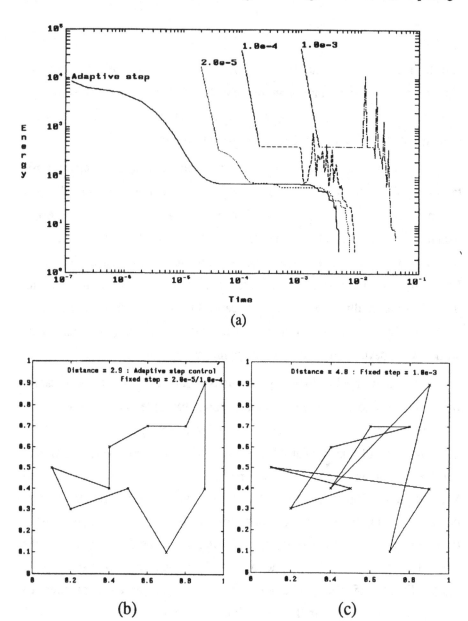

Fig. 1.5 Network dynamic of a 10-city TSP.
    (a) Transient behavior.
    (b) Solution with adaptive time-step and with fixed time-step
        of 1.0×10^{-4} and 1.0×10^{-5}.
    (c) Solution with fixed time-step of 1.0×10^{-3}.

This result is exactly contradictory to the original Hopfield theory that the network always operates toward the direction of decreasing energy function. This example therefore shows that the asynchronous operation has an advantage over the synchronous operation in terms of computational time and solution accuracy.

Several research projects address the issues of novel placement and routing of neuron and synapse cells [21]. One such activity is to design a silicon retina and cochlea [22]. MOS transistors in these chips operate in the subthreshold region to achieve high voltage gain and low power dissipation. These analog VLSI chips perform the sensory functions with preliminary signal processing similar to human organs. These electronic neural networks can be characterized with high locality that one neuron is tied to just the neighboring neurons. The input signals are processed locally and gradually spread to the whole network in a decayed format. The high locality property is seldom explored in other neural networks.

One neural network hardware designed for signal processing purpose is the VLSI chip from Bell Laboratories [17]. The chip has 256 neurons and 100,000 synapses precoded during the fabrication stage. The total chip size is 5700 μm x 5700 μm in a 2.5-μm CMOS technology. The weightings of synapses, which are made from amorphous-Si, are defined by electron beams during wafer fabrication. It is known that the main application of this chip is to compress the bandwidth of video image for telephone transmission. Even though the synapse cell is very compact, the synapse weighting is fixed at the fabrication step. Another example to obtain high resistivity material for synapse weightings in the standard CMOS technology is to use well resistors [18].

A limited programmability for a content-addressable memory can be obtained by using switching MOS transistors [23]. The on/off switches, which are controlled by data stored in the on-chip memory, determine the fixed synapse weight. The synapses have only two states: inhibitory and excitatory. Each chip has 54 neurons and 2916 synapses and occupies 6700 μm x 6700 μm area in a 2.5-μm CMOS technology. More synapse weighting selections can be made by switching certain ratioed current mirrors with the current steering structure [24].

An analog multiplier with EEPROM cells is used to build a fully programmable synapse [25]. The synapse weighting is decided by device aspect ratios and the bias current. Since the EEPROM-injected electronic charge at the floating gate can change the threshold voltage of the transistor in the current mirror circuitry, bias currents of the analog multiplier are programmable. However, the voltage range V_{IR} for a linear conductance is limited by the smaller dynamic range of the two differential pairs. For a proper input voltage range, the device W/L ratio should be very small. A similar circuit cell with DRAM cells is also developed [26]. Due to continuous charge leakage in the DRAM cell, periodic refreshing is needed. MOS transistors are used to store the synapse weighting information.

The digitally programmable synapse weighting can be made by controlling the RC charging time [27]. The synapse weighting data stored in the digital memory are converted into the switching pulse width. In this approach, the charge summation and neural decision are done in the analog fashion. The fully digital neural chip includes an on-chip random signal generator [28]. The synapse weighting is coded into the pulse width and the summation function in neuron is realized with the AND logic gates. Since the mark time and occurrence on the pulse train is determined by a random number generator, such a sigmoid function in the neuron is performed by a stochastic process. In this architecture, the dynamic range of the synapse weighting is limited by the resolution of the pulse width so that a very long processing time is necessary for the applications with a large ratioed synapse weighting

[29].

In order to explore the rich properties of neural networks associated with massively parallel processing using analog neurons and synapses, compact and programmable VLSI neural circuitries are needed.

References

[1] R. P. Lippman, "An introduction to computing with neural nets," *IEEE Acoustics, Speech, and Signal processing Magazine*, pp. 4-22, April 1987.

[2] R. Hecht-Nielsen, "Neural-computing: picking the human brain," *IEEE Spectrum*, vol. 25, no. 3, pp. 36-41, Mar. 1988.

[3] J. A. Anderson and E. Rosenfeld, *Neuralcomputing - Foundation of Research*, Cambridge, MA: The MIT Press, 1988.

[4] S. Grossberg, *Neural Network and Natural Intelligence*, Cambridge, MA: The MIT Press, 1988.

[5] W. S. McCulloch, W. Pitts, "A logical calculus of the idea immanent in neural nets," *Bulletin of Mathematical Biophysics*, vol. 5, pp. 115-133, 1943.

[6] F. Rosenblatt, *Principles of neurodynamics: perceptrons and the theory of brain mechanisms*, Spartan Books, Washington D.C., 1961.

[7] B. Widrow, Bernard, Hoff, and Marcian, "Adaptive switching circuits," *1960 IRE WESCON Convention Record*, Part 4, pp. 96-104, Aug. 23-26, 1960.

[8]   J. A. Anderson, "A simple neural network generating an interactive memory," *Mathematical Biosciences,* vol. 14, pp. 197-220, 1972.

[9]   J. J. Hopfield, "Neural network and physical systems with emergent collective computational abilities," *Proc. Natl. Acad., Sci. U.S.A.,* vol. 79, pp. 2554-2558, Apr. 1982.

[10] B. Kosko, "Adaptive bidirectional associative memories," *Applied Optics,* vol. 36, pp. 4947-4960, Dec. 1987.

[11] D. Hebb, *The Organization of Behavior,* New York: Wiley, 1949.

[12] T. Kohonen, *Self-Organization and Associative Memory,* 2nd Ed., New York: Springer-Verlag, 1987.

[13] R. Hecht-Nielsen, "Counter-propagation networks," *Proc. of IEEE First Inter. Conf. on Neural Networks,* vol. II, pp. 19-32, San Diego, CA, 1987.

[14] A. H. Klopf, "A drive-reinforcement model of single neuron function: an alternative to the Hebbian neural model," *Proc. of Conf. on Neural Networks for Computing,* pp. 265-270, Snowbird, UT, Apr. 1986.

[15] G. E. Hinton and T. J. Sejnowski, "A learning algorithm for Boltzmann machines," *Cognitive Science,* vol. 9, pp. 147-169, 1985.

[16] Y. P. Tsividis, "Analog MOS integrated circuits - certain new ideas, trends, and obstacles," *IEEE Jour. of Solid-State Circuits,* vol. SC-22, no. 3, pp. 317-321, June 1987.

[17] H. P. Graf and P. deVegvar, "A CMOS implementation of a neural network model," *Proc. of the Stanford Advanced Research in VLSI Conference,* pp. 351-362, Cambridge, MA: The MIT

Press, 1987.

[18] B. W. Lee and B. J. Sheu, "Design of a neural-based A/D con-
      verter using modified Hopfield network," *IEEE Jour. of Solid-State
      Circuits,* vol. SC-24, no. 4, pp. 1129-1135, Aug. 1989.

[19] Dahlquist, Björck, and Anderson, *Numerical Methods,* pp. 269-273,
      Englewood Cliffs, NJ: Prentice-Hall, 1974.

[20] J. J. Hopfield, "Neurons with graded response have collective com-
      putational properties like those of two-state neurons," *Proc. Natl.
      Acad., Sci. U.S.A.,* vol. 81, pp. 3088-3092, May 1984.

[21] P. Treleaven, M. Pacheco, and M. Vellasco, "VLSI architectures
      for neural networks," *IEEE Micro Magazine,* vol. 9, no. 6, pp. 8-
      27, Dec. 1989.

[22] C. A. Mead, *Analog VLSI and Neural Systems,* New York:
      Addison-Wesley, 1989.

[23] R. E. Howard, D. B. Schwartz, J. S. Denker, R. W. Epworth, H.
      P. Graf, W. E. Hubbard, L. D. Jackel, B. L. Straughn, and D. M.
      Tennant, "An associative memory based on an electronic neural
      network architecture," *IEEE Trans. on Electron Devices,* vol. ED-
      34, no. 7, pp. 1553-1556, July 1987.

[24] P. Mueller, J. V. D. Spiegel, D. Blackman, T. Chiu, T. Clare, C.
      Donham, T. P. Hsieh, M. Loinaz, "Design and fabrication of VLSI
      components for a general purpose analog neural computer," in
      Analog VLSI Implementation of Neural Systems, Editors: C. Mead
      and M. Ismail, Boston, MA: Kluwer Academic, pp. 135-169, 1989.

[25] M. Holler, S. Tam, H. Castro, R. Benson, "An electrically train-
      able artificial neural network (ETANN) with 10240 'float gate'

synapses," *Proc. of IEEE/INNS Inter. Joint Conf. on Neural Networks,* vol. II, pp. 191-196, Washington D.C., June 1989.

[26] T. Morishita, Y. Tamura, and T. Otsuki, "A BiCMOS analog neural network with dynamically updated weights," *Tech. Digest of IEEE Inter. Solid-State Circuits Conf.,* pp. 142-143, San Fransisco, CA, Feb. 1990.

[27] A. F. Murray, "Pulse arithmetic in VLSI neural network," *IEEE Micro Magazine,* vol. 9, no. 6, pp. 64-74, Dec. 1989.

[28] D. E. Van den Bout and T. K. Miller III, "A digital architecture employing stochasticism for the simulation of Hopfield neural nets," *IEEE Trans. on Circuits and Systems,* vol. 36, no. 5, pp. 732-746, May 1989.

[29] M. S. Tomlinson Jr., D. J. Walker, M. A. Sivilotti, "A digital neural network architecture for VLSI," *Proc. of IEEE/INNS Inter. Joint Conf. on Neural Networks,* vol. II, pp. 545-550, San Diego, CA, June 1990.

Chapter 2

VLSI Hopfield Networks

The Hopfield networks [1] are very popular for electronic neuro-computing due to the simplicity in the network architecture and the fast convergence property. A Hopfield network composed of one-layer neurons and fully connected feedback synapses can be used to realize associative memory, pattern classifier, and optimization circuits. The network always operates along a decreasing direction for the energy function, so that the final output represents one minimum of the energy function. Due to the complexity of energy function, there could exist several local minima. The minima in the energy function of Hopfield networks, which are encoded in the resistive network, are used for the exemplar patterns in associative memory and pattern classifier applications. However, the existence of local minima is not desirable for a great variety of other optimization applications.

Another major reason Hopfield neural networks are widely used for VLSI implementation is the relaxed requirement on network elements. The key requirement on neuron transfer function for the Hopfield network is to achieve monotonic functions, while that for other neural networks is usually an exact mathematical expression like sigmoid, Boltzmann, or Gaussian function. Since the amplifier gain in analog VLSI neural circuits cannot follow an exact function, the simpler neuron transfer function requirement makes Hopfield networks a popular choice for VLSI implementation. The rich local minima can help to operate without severe performance degradation in a situation where the device variations are large.

Hopfield networks can be used as an effective interface between the real world of analog transmission media and the digital computing machines. Inputs to the Hopfield network can be analog signals. Outputs are usually discrete values as shown in Fig. 2.1. The Hopfield network not only converts analog signals into the digital format but also can conduct the first-level signal processing, such as associative recalling, signal estimation, and combinatorial optimization in a way similar to the human retina. Due to the robustness of neural processors, output of the Hopfield neural network need not be done with high precision, which is in strong contrast to conventional interface circuits.   Key advantages of

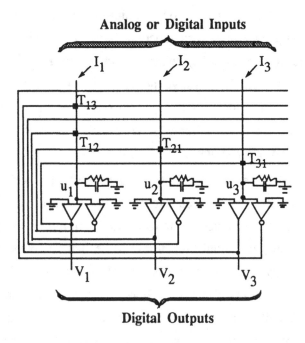

Fig. 2.1 A Hopfield neural network with inverting
and non-inverting amplifiers as neurons.

the VLSI neural interfaces over conventional interfaces are derived from the learning capabilities of Hopfield networks [2]. By adjusting conductance values between the amplifiers with a learning rule, adaptive characteristics can be made. The adaptability of a neural-based A/D converter [3], for example, will be useful not only to compensate for initial device mismatches or long-term characteristic drifts but also to provide a greater processing capability in the image and signal processing system [4-6].

2.1 Circuit Dynamics of Hopfield Networks

In a Hopfield network where the resistive network T_{ij} is symmetric and without self-feedback terms, i.e. $T_{ij} = T_{ji}$ and $T_{ii} = 0$, the energy function E can be expressed as

$$E = -\frac{1}{2}\sum_{i=1}^{n}\sum_{j=1, j\neq i}^{n} T_{ij}V_iV_j - \sum_{i=1}^{n} I_iV_i + \sum_{i=1}^{n} T_i \int_0^{V_i} g_i^{-1}(V)\, dV \ . \ (2.1)$$

Here, T_{ij} is the conductance between the i-th amplifier input and the j-th amplifier output, V_i is the i-th amplifier output voltage, I_i is the input current signal to the i-th amplifier, $g(\ . \ )$ is the amplifier transfer function, and n is the number of neurons. By using the energy function, transient behavior of the Hopfield network can be explained in the following way. Input voltage of the i-th amplifier u_i is governed by Kirchoff's Current Law,

$$C_i \frac{du_i}{dt} = \sum_{j=1, j\neq i}^{n} T_{ij}V_j - T_iu_i + I_i \ . \qquad (2.2)$$

Here, C_i and T_i are the capacitance and the total conductance at the i-th amplifier input node, respectively. The total input conductance for each amplifier, T_i, is given as

$$T_i = \sum_{j=1, j\neq i}^{n} |T_{ij}| \ . \qquad (2.3)$$

In the initial analog VLSI circuit realization of a Hopfield network, simple decision-making amplifiers and a resistive network are used. Each neuron includes noninverting and inverting amplifiers to handle both positive and negative synapse weightings. The output levels of a noninverting amplifier are 0 V and 1 V, while that of an inverting amplifier are 0 V and -1 V. The amplifier outputs are fed back to the amplifier inputs through the densely connected resistive network. The RC charging effect from the T_i and C_i in the original Hopfield network is equivalently obtained with the finite bandwidth of the amplifier in the VLSI implementation.

The time derivative of the energy function can be determined as

$$\frac{dE}{dt} = - \sum_{i=1}^{n} \frac{dV_i}{dt} \left[\sum_{j=1, \, j \neq i}^{n} T_{ij} V_j - T_i u_i + I_i \right]. \tag{2.4}$$

Substituting (2.2) into (2.4), we obtain

$$\frac{dE}{dt} = - \sum_{i=1}^{n} \frac{dV_i}{dt} \, C_i \, \frac{du_i}{dt}$$

$$= - \sum_{i=1}^{n} C_i g_i^{-1}(V_i) \left[\frac{dV_i}{dt} \right]^2. \tag{2.5}$$

If $g_i(\,.\,)$ is monotonically increasing, $\frac{dE}{dt}$ is always negative. That is to say, the network moves in the direction of decreasing the energy function. When $\frac{dV_i}{dt} = 0$ for all i, the steady state is reached. Due to the nature of the energy function, the network output is highly dependent upon its initial state. Thus, the final output is usually the closest local minimum from the initial state of the network.

The energy function can be used to describe the macroscopic property of network dynamics. However, it is insufficient to explain such detailed behaviors of VLSI Hopfield circuits as the location of local minima, convergence speed, and required amplifier gain for proper network operation.

2.2 Existence of Local Minima

Detailed properties of the local minima can be understood with the help of the following analysis method [7]. At a stable output, the input voltage to each neuron is governed by

$$
\begin{cases}
u_i > 0 & \text{when } V_i = 1\ V \\
u_i < 0 & \text{when } V_i = 0\ V\ .
\end{cases}
\tag{2.6}
$$

Here, the input voltage of the i-th neuron, u_i, is determined by Kirchoff's Current Law,

$$
C_i \frac{du_i}{dt} + T_i u_i = \sum_{j \ne i,\ j=1}^{n} T_{ij} V_j + I_i\ .
\tag{2.7}
$$

From (2.6), the range of input current I_i for each stable state can be calculated,

$$
\begin{cases}
I_i > - \displaystyle\sum_{j \ne i,\ j=1}^{n} T_{ij} V_j & \text{for } V_i = 1\ V \\
I_i < - \displaystyle\sum_{j \ne i,\ j=1}^{n} T_{ij} V_j & \text{for } V_i = 0\ V\ .
\end{cases}
\tag{2.8}
$$

During the transient period, the voltage V_i is changing in the direction that the energy function E decreases. When all amplifiers reach a stable condition in (2.6), the searching process is terminated. From (2.8), a characteristic parameter GAP can be defined as the input current difference of the lower limit for $V_i = 1\ V$ and the upper limit for $V_i = 0\ V$,

$$
GAP_i \equiv - \sum_{j \ne i,\ j=1}^{n} T_{ij}(V_j^\mu - V_j^l)\ ,
\tag{2.9}
$$

where V_j^u and V_j^l are the output voltages of the digital codes to the j-th amplifier output whose i-th bit from the least significant bit (LSB) are logic 1 and logic 0, respectively. Here, the unit of GAP is ampere. Since the parameter GAP_i represents the overlapped range of the i-th input current, two digital codes can be stable for a given input current when $GAP_i < 0$ for every i as shown in Fig. 2.2. Therefore, the condition when the two digital codes cannot be local minima in the energy function for a given analog input value is $GAP_i \geq 0$ for every i.

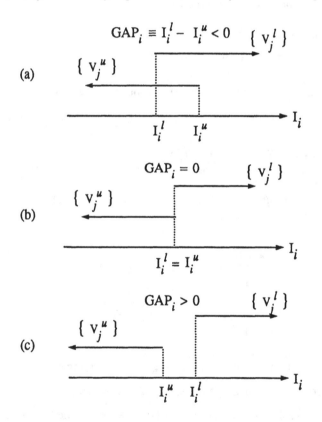

Fig. 2.2 Calculation of a characteristic parameter GAP.

        (a) $GAP_i < 0$.

        (b) $GAP_i = 0$.

        (c) $GAP_i > 0$.

Let us examine the special case that $\{ V_i^\mu \}$ and $\{ V_i^l \}$ are adjacent digital codes. The parameter GAP_k can be simplified as

$$GAP_k = \begin{cases} \displaystyle\sum_{j=1}^{k-1} T_{kj} & \text{if } k > 1 \,, \\ 0 & \text{if } k = 1 \,. \end{cases} \qquad (2.10)$$

Here, $\{ V_i^\mu \}$ and $\{ V_i^l \}$ are

$$\begin{cases} V_i^\mu = V_i^l & \text{if } i > k \\ V_i^\mu = 1 \ V \quad \text{and} \quad V_i^l = 0 \ V & \text{if } i = k \\ V_i^\mu = 0 \ V \quad \text{and} \quad V_i^l = 1 \ V & \text{if } i < k \,. \end{cases} \qquad (2.11)$$

For a set of adjacent digital codes with the only different bit occurring in the LSB, the codes cannot be minima at the same input. Even though $\displaystyle\sum_{j=1}^{k-1} T_{ij} \geq 0$ is just a necessary condition that both the adjacent codes cannot be stable at a given input, it can be used as an indicator for the existence of local minima. For example, all feedback synapse weightings of the Hopfield neural-based A/D converter [3] are negative, so that the necessary condition is not satisfied.

In a general case when two digital codes have different values at the k-th bit but have the same value above the k-th bit, the parameter GAP_i can be obtained as follows,

$$GAP_i = \begin{cases} \text{undefined} & \text{if } i > k \,, \\ \displaystyle\sum_{j=1}^{k-1} T_{kj}(V_j^\mu - V_j^l) & \text{if } i = k \,, \quad (2.12) \\ (V_i^l - V_i^\mu)\left[T_{ki} + \displaystyle\sum_{j \neq i,\, j=1}^{k-1} T_{ij}(V_j^\mu - V_j^l) \right] & \text{if } i < k \,. \end{cases}$$

Here, the digital codes are given as

$$\begin{cases} V_j^u = 1 \ V \ \text{and} \ V_j^l = 0 \ V & \text{for} \ j = k \\ V_j^u = V_j^l & \text{for} \ j > k \ . \end{cases} \tag{2.13}$$

The condition that two digital states can be stable at an input current is

$$GAP_i < 0 \qquad \text{for every} \ i \le k \ . \tag{2.14}$$

2.3 Elimination of Local Minima

The overlapped input range between two digital codes can be eliminated by adding correction current $\{I_{iC}\}$ into the neuron inputs as shown in Fig. 2.3. The correction logic gates monitor the Hopfield network outputs and generate the correcting information. The correction currents are fed back into the neuron inputs through the extended conductance network. Here, the response time of the correction circuits is smaller than that for the rest of the network. The input voltage u_i with the correcting current is governed by

$$T_i u_i = \sum_{j \ne i, \ j=1}^{n} T_{ij} V_j + I_i + I_{iC} \ . \tag{2.15}$$

Therefore, the parameter GAP_i in (2.12) is changed to GAP_i^C as follows,

$$GAP_i^C = \begin{cases} \displaystyle\sum_{j=1}^{k-1} T_{kj}(V_j^u - V_j^l) - I_{kC}^u + I_{kC}^l & \text{for} \ i = k \\ (V_i^l - V_i^u)\left[\displaystyle\sum_{j=1}^{k-1} T_{ij}(V_j^u - V_j^l) + T_{ki} + I_{iC}^u - I_{iC}^l\right] + T_{ii} & \text{for} \ i < k \ . \end{cases} \tag{2.16}$$

The new energy function E_C is given as

$$E_C = -\frac{1}{2} \sum_{i=1}^{n} \sum_{j \ne i, \ j=1}^{n} T_{ij} V_i V_j - \sum_{i=1}^{n} (I_i + I_{iC}) V_i$$

$$= E_O - \sum_{i=1}^{n} I_{iC} V_i. \tag{2.17}$$

Here, E_O and E_C are the original and modified energy functions, respectively. The additional term in the modified energy function is introduced to fill up the wells of local minima, so that only the global minimum exists.

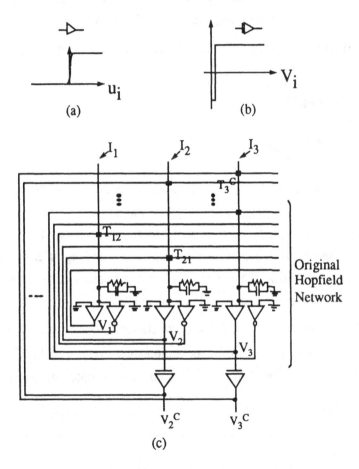

Fig. 2.3 Modified Hopfield neural network without local minima.
    (a) Transfer characteristics of the Hopfield amplifier.
    (b) Transfer characteristics of the correction amplifier.
    (c) Modified Hopfield neural network.

The general criteria for choosing the correction current to eliminate local minima in Hopfield networks are described below.

Criterion-A: Avoid the condition that $GAP_i^C < 0$ for every $i < k$, which makes both the digital codes in (2.13) stable.

Criterion-B: Preserve network dynamics of Hopfield networks so that the modified network moves in the direction of decreasing the energy function.

Criterion-C: Preserve the input current range for a global minimum.

Criterion-B requires that

$$\sum_{i=1}^{n} V_i \frac{\partial I_{iC}}{\partial V_i} = 0 .$$ (2.18)

Notice that criterion-C is often difficult to satisfy in some optimization problems with no *a priori* knowledge of the solutions. For example, the solution of a Hopfield neural-based A/D converter at a given analog input voltage is well defined, while that of a traveling salesman problem is not easy to know. The choices of the correction current for the two distinct cases will be described later in detail.

From criterion-A, $GAP_k^C = 0$ is chosen. Notice that the requirement in (2.14) is logic-AND condition for every $i < k$. By setting $I_{kC}^u = - I_{kC}^l$, the correction currents at the steady state are given as

$$I_{kC}^u = - I_{kC}^l = T_{kC} ,$$ (2.19)

where

$$T_{kC} = \frac{1}{2} \sum_{j=1}^{k-1} T_{kj} (V_j^u - V_j^l) .$$ (2.20)

The correcting currents during the searching process can be expressed as

$$I_{kC} \equiv I_{kC}^l - I_{kC}^u$$

$$= T_{kC} f^u(V_k) + T_{kC} f^l(V_k) \ . \tag{2.21}$$

Here, the boundary conditions of $f^u(V_k)$ and $f^l(V_k)$ are determined from the steady-state condition in (2.13) as follows,

$$\begin{cases} f^u(0) = 0, f^u(1) = 1 \ ; \text{ and} \\ f^l(0) = 1, f^l(1) = 0 \ . \end{cases} \tag{2.22}$$

Functions $f^u(V_k)$ and $f^l(V_k)$ can be determined by criterion-B,

$$V_k \frac{\partial I_{kC}}{\partial V_k} = T_{iC} V_k \left[\frac{\partial f^u(V_k)}{\partial V_k} - \frac{\partial f^l(V_k)}{\partial V_k} \right] = 0 \ . \tag{2.23}$$

If the functions are selected as

$$f^u(V) = U(V + \varepsilon) \tag{2.24}$$

and

$$f^l(V) = U(\varepsilon) - U(V + \varepsilon) \tag{2.25}$$

then, (2.23) becomes

$$V_k \frac{\partial I_{kC}}{\partial V_k} = T_{kC} 2V_k \delta(V_k + \varepsilon) \approx 0 \ . \tag{2.26}$$

Here, $U(V)$ is a step function, $\delta(V)$ is a delta function, and ε is a very small positive number. Notice that the above selection is made by decreasing the input overlapped range by half from each side of the two digital codes.

In the following sections, the self-correction techniques are applied to eliminate the local minima for the neural-based A/D conversion and traveling salesman problems. The analog-to-digital conversion is a simple and straightforward example to demonstrate the capabilities of the modified Hopfield networks because the optimal solution is clearly known. The traveling salesman problem is with the intrinsically unknown global minimum. Our modified Hopfield networks find excellent solutions to both practical problems.

2.4 Neural-Based A/D Converter Without Local Minima

The function of an A/D converter is to find a digital word $\{V_n V_{n-1} \cdots V_1\}$ which is a best representation of the analog input signal V_S. The output of an A/D converter is to achieve the minimum of the following function

$$E_o = (V_S - \sum_{i=1}^{n} V_i 2^{i-1})^2 .$$ (2.27)

In order to construct a Hopfield-like energy function, some modifications have to be made,

$$E = \frac{1}{2} E_o - \frac{1}{2} \sum_{i=1}^{n} (2^{i-1})^2 [V_i (V_i - 1)] .$$ (2.28)

Notice that the additional term, which eliminates the diagonal elements in the Hopfield network, does not affect the final correct solution [3]. By expanding (2.28), the objective function for the A/D converter can be expressed as

$$E = - \frac{1}{2} \sum_{i=1}^{n} \sum_{j=1, \ j \neq i}^{n} (-2^{i+j-2}) V_i V_j - \sum_{i=1}^{n} (-2^{2i-3} + 2^{i-1} V_S) V_i .$$

(2.29)

The synapse weightings T_{ij} and input current I_i can be determined from (2.1) and (2.29),

$$T_{ij} = -2^{i+j-2}$$ (2.30)

and

$$I_i = -2^{2i-3} + 2^{i-1} V_S .$$ (2.31)

The circuit schematic of a 4-bit A/D converter is shown in Fig. 2.4. The input current I_i is also implemented by using the conductances (T_{iR} and T_{iS}) and the reference voltage $V_R = -1$ V. Since all actual conductance values are positive in the VLSI implementation, the polarity for output voltage V_i is changed to [−1 V, 0 V] by using inverting amplifiers. The values for the conductances in the A/D converter are

$$G_{ij} = 2^{i+j-2}, \tag{2.32}$$

$$G_{iR} = 2^{2i-3}, \tag{2.33}$$

and

$$G_{iS} = 2^{i-1}. \tag{2.34}$$

Here, G_{iR} is the conductance between the input terminal of the i-th amplifier and the reference voltage V_R, and G_{iS} is the conductance bet-

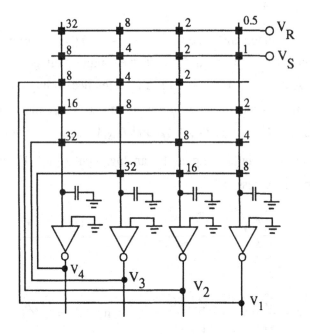

Fig. 2.4 A 4-bit neural-based A/D converter.

ween the input terminal of the i-th amplifier and the analog input voltage V_S. Notice that T and G are used interchangeably to represent the conductances of the network elements.

The amplifier input voltage $\{u_i\}$ is governed by

$$\left[\sum_{j=1,\, j\neq i}^{n} G_{ij} + G_{iR} + G_{iS} \right] u_i = \sum_{j=1,\, j\neq i}^{n} G_{ij} V_j + G_{iS} V_S + G_{iR} V_R \quad (2.35)$$

Since the input voltage $\{u_i\}$ is determined by the ratios of conductances, the scaling factor to realize absolute conductance values can be used as an integrated-circuit design parameter. The normalized conductances are

$$\hat{G}_{iS} = 1 , \tag{2.36}$$

$$\hat{G}_{ij} = 2^{j-1} , \tag{2.37}$$

and

$$\hat{G}_{iR} = 2^{i-2} . \tag{2.38}$$

The maximum conductance ratio for an n-bit A/D converter has been greatly reduced from 2^{2n-2} to 2^n. If the output voltage range of the amplifier is scaled to $(0 \text{ V}, -V_A)$, the reference voltage is scaled to $-V_{ref}$, and the conversion step size is scaled to V_{step}, then the conductances are scaled to $\hat{G}_{ij}/V_A$, $\hat{G}_{iR}/V_{ref}$ and $\hat{G}_{iS}/V_{step}$, respectively.

For a specific digital output, the analog input signal V_S to the i-th amplifier can be expressed as

$$V_S > - \frac{\hat{G}_{iR}}{\hat{G}_{iS}} V_R - \sum_{j=1,\, j\neq i}^{n} \frac{\hat{G}_{ij}}{\hat{G}_{iS}} V_j \quad \text{when } V_i = -1 \text{ V} \tag{2.39}$$

and

$$V_S < - \frac{\hat{G}_{iR}}{\hat{G}_{iS}} V_R - \sum_{j=1,\, j\neq i}^{n} \frac{\hat{G}_{ij}}{\hat{G}_{iS}} V_j \quad \text{when } V_i = 0 \text{ V} . \tag{2.40}$$

Since

$$-\frac{\hat{G}_{iR}}{\hat{G}_{iS}}V_R - \sum_{j=1,\ j\neq i}^{n}\frac{\hat{G}_{ij}}{\hat{G}_{iS}}V_j = 2^{i-2} + V_O - 2^{i-1}|V_i|\ , \tag{2.41}$$

the upper limit and lower limit of the analog input voltage always increase with i. Here, $V_R = -1\ V$ and V_O is the digital output voltage which equals to $\sum_{i=1}^{n}2^{i-1}|V_i|$. To achieve a stable output, the input signal range is governed by the logic-AND operation of the range decided by each amplifier. Therefore, the lower limit of V_S at a given digital output is determined by the first logic-1 occurrence of the digital word from the least significant bit (LSB) and the upper limit is determined by the first logic-0 occurrence from the LSB, as shown in Table 2.1. If a digital word has the first logic-0 at the i-th bit from the LSB, then the next adjacent digital word has the first logic-1 at the i-th bit. Thus, the i-th amplifier decides the upper limit and lower limit of the analog input voltage between two adjacent digital words. That is to say, if the upper limit of the input signal corresponding to a digital word is decided by the i-th amplifier, the lower limit of the upper adjacent digital word is also decided by the i-th amplifier.

To guarantee that only one global minimum corresponding to each analog input value exists (i.e., without local minima), only one-to-one correspondence between the digital output and analog input should occur. The characteristic parameter GAP can be used to understand the existence of local minima. It can be expressed as

$$GAP = -\sum_{j=1}^{k-1}2^{j-1}(\ V_j^l - V_j^u\ ) \tag{2.42}$$

where $\{V_j^l\}$ and $\{V_j^u\}$ are the adjacent digital words in (2.11). Since input currents for the neural-based A/D converter are determined by the analog input voltage, only one characteristic parameter exists. If the GAP is negative, both the adjacent digital words can act as the stable digital output at a given analog input voltage. However, if the *GAP* is positive, there can be a gap of analog input voltage where no stable digital output exists. For a proper A/D conversion, the *GAP* should be zero.

Table 2.1 Analog input range for each digital code

DIGITAL CODE	ANALOG INPUT RANGE			
D_4 D_3 D_2 D_1	V_1	V_2	V_3	V_4
0 0 0 0	< 0.5	< 1.0	< 2.0	< 4.0
0 0 0 1	> 0.5	< 2.0	< 3.0	< 5.0
0 0 1 0	< 2.5	> 1.0	< 4.0	< 6.0
0 0 1 1	> 2.5	> 2.0	< 5.0	< 7.0
0 1 0 0	< 4.5	< 5.0	> 2.0	< 8.0
0 1 0 1	> 4.5	< 6.0	> 3.0	< 9.0
0 1 1 0	< 6.5	> 5.0	> 4.0	< 10.0
0 1 1 1	> 6.5	> 6.0	> 5.0	< 11.0
1 0 0 0	< 8.5	< 9.0	< 10.0	> 4.0
1 0 0 1	> 8.5	< 10.0	< 11.0	> 5.0
1 0 1 0	< 10.5	> 9.0	< 12.0	> 6.0
1 0 1 1	> 10.5	> 10.0	< 13.0	> 7.0
1 1 0 0	< 12.5	< 13.0	> 10.0	> 8.0
1 1 0 1	> 12.5	< 14.0	> 11.0	> 9.0
1 1 1 0	< 14.5	> 13.0	> 12.0	> 10.0
1 1 1 1	> 14.5	> 14.0	> 13.0	> 11.0

The characteristic parameter *GAP* becomes

$$GAP = -2^{k-1} + 1 . \qquad (2.43)$$

The above equation shows that the *GAP* is always negative except when

$k = 1$. Hence, there could exist more than two digital output words corresponding to a given analog input value. In addition, the overlapped length is not the same for every output word in the Hopfield A/D converter. Figure 2.5 shows the input signal range corresponding to each digital word. The overlapped digital words for a given analog input value are caused by the local minima in the energy function. The largest overlap in the analog input voltage occurs at the words (0111) and (1000) where the 4-th amplifier decides the analog input range. This phenomenon is apparent because the characteristic parameter *GAP* increases with k. However, there is no overlapped input range between the adjacent digital words if these two words differ by the LSB.

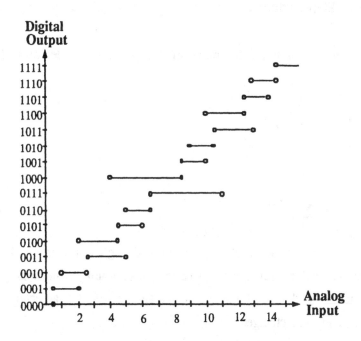

Fig. 2.5 Digital output versus analog input characteristics of
the A/D converter shown in Fig. 2.4.

The overlapped regions in Fig. 2.5 indicate the existence of local minima in the Hopfield neural-based A/D converter proposed by Tank and Hopfield [3]. For example, the words (1000), (0110), and (0101) can be the stable outputs when $V_S = 5.6\ V$. The converged output in analog VLSI chips is determined by the initial state of the network and device mismatches. The local minima give severe nonideal output characteristics. While nonlinearities can be remedied in software computation to some extent by resetting the network to the ground state prior to each conversion cycle, the circuit transfer characteristic in VLSI chips still has strong nonlinearity due to amplifier mismatches.

2.4.1 The Step Function Approach

The analog input voltage range of a digital code is determined as

$$\begin{cases} V_S \geq V_o - 2^{p-2} \\ V_S \leq V_o + 2^{q-2}, \end{cases} \tag{2.44}$$

where

$$V_o = \sum_{i=1}^{n} 2^{i-1} V_i . \tag{2.45}$$

Here, p and q are the first logic-1 bit and logic-0 bit from the LSB, respectively. After adding the correction current terms, the analog input voltage range and the characteristic parameter of the adjacent digital codes in (2.11) are changed as

$$\begin{cases} V_S^C \geq V_o^l - 2^{k-2} - \dfrac{T_{kC}}{2^{k-1}}\, f^l(V_k^l) \\[4mm] V_S^C \leq V_o^u + 2^{k-2} - \dfrac{T_{kC}}{2^{k-1}}\, f^u(V_k^u) \end{cases} \tag{2.46}$$

and

$$GAP^C = -2^{k-1} - 1 - \frac{T_{kC}}{2^{k-1}} \left[f^l(V_k^l) - f^u(V_k^u) \right] . \tag{2.47}$$

Here, V_S^C and GAP^C are the corrected input voltage and characteristic parameter, respectively. Since analog input ranges of a neural-based A/D converter without local minima are given as

$$V_o - 0.5 < V_S < V_o + 0.5 , \tag{2.48}$$

the analog input range should be corrected to be at the midpoint of the overlapped range. Notice that the choice of $\{T_{iC}\}$ for the neural-based A/D converter can keep the analog input voltage range of global minimum, because the voltage range can be well defined at a given analog input. Therefore, the correction current for the codes is

$$I_{kC} = -\frac{2^{k-1} - 1}{2} \left[f^u(V_k) - f^l(V_k) \right] . \tag{2.49}$$

Here, $f^u(V)$ and $f^l(V)$ are given in (2.24) and (2.25).

The energy function of a Hopfield neural-based A/D converter is shown in Fig. 2.6. The original Hopfield energy function has two local minima at $V_S = 1.8\ V$, which are (0001) and (0010). Due to the two energy wells, the final converged output is determined by the initial states of the network. Notice that the two digital codes are the local minima in the transfer curve shown in Fig. 2.5. The corrected energy surface shown in Fig. 2.7 has only one minimum. Therefore, the final output is always the global minimum. Here, the correction functions used in this simulation are

$$f^u(x) = \frac{1}{1 + \exp(-100000 \times (x - 0.001))} \tag{2.50}$$

and

$$f^l(x) = \frac{1}{1 + \exp(-100000 \times (0.999 - x))} . \tag{2.51}$$

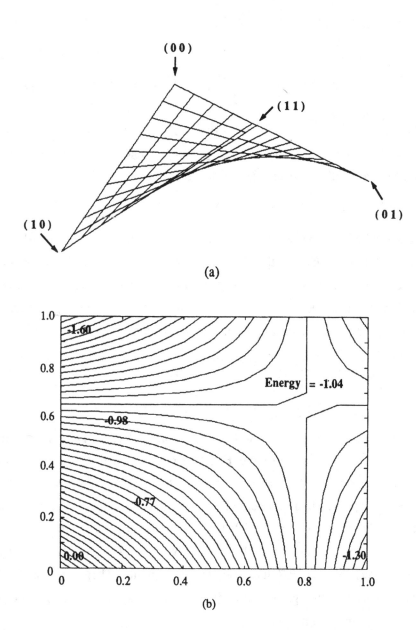

Fig. 2.6   Energy function of the original Hopfield A/D converter.
(a) Energy surface with changing V_1 and V_2.
(b) Contour of (a).

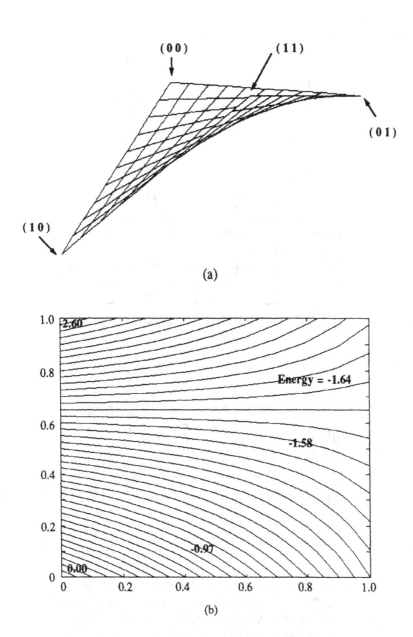

(a)

(b)

Fig. 2.7   Energy function of a modified Hopfield A/D converter.
         (a) Energy surface with changing V_1 and V_2.
         (b) Contour of (a).

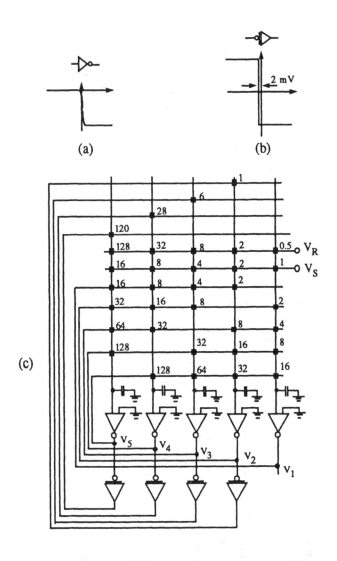

Fig. 2.8 A 5-bit modified neural -based A/D converter.
        (a) Transfer characteristics of the Hopfield amplifier.
        (b) Transfer characteristics of the correction amplifier.
        (c) Complete circuit diagram

Figure 2.8 shows the circuit schematic of a neural-based A/D converter with the correction technique. In the hardware experiments, standard integrated-circuit parts were used to construct the 5-bit Hopfield

neural-based A/D converter. Amplifiers function as electronic neurons and resistors function as electronic synapses. To realize the negative synapse weighting, the amplifier output polarity is reversed. Voltage comparators are used to build the correction circuit, whose reference voltage is a very small negative value of −2 mV. The measured voltage transfer curves are shown in Fig. 2.9. With the correcting currents, the nonlinearities in the Hopfield A/D converter caused by the local minima are completely removed.

2.4.2 The Correction Logic Approach

Two design problems exist in the correction method described in section 2.4.1. First, the previous correction is obtained from violating the local minima existence condition that $GAP_k < 0$ of the digital codes in (2.13). Table 2.2 shows the analog input range after the previous correcting scheme is applied. Since the corrected input ranges for each neuron are the same, all correcting currents for a digital code should be exactly equal. In the VLSI implementation, this precision is difficult to achieve. Second, the offset voltage of the correction amplifier should be very small in order to keep the same network dynamics as the Hopfield neural network. In VLSI technologies, such a small and uniform offset voltage through all correction amplifiers is very difficult to realize.

The second approach to eliminate the local minima in the Hopfield neural-based A/D converter avoids these two design problems. Since the overlapped input range for the neural-based A/D converter is determined by the adjacent digital codes as described in the previous section, a detailed correction can be conducted between the specific codes [5]. The correction logic circuitry is used to generate the correction current at a specific digital code. Truth table of the correction logic is listed in Table 2.3. The output of the correction logic circuitry can take a

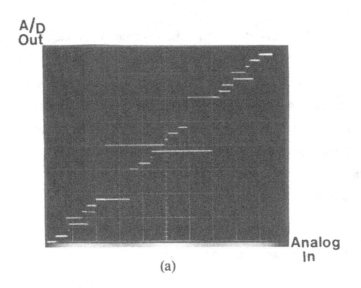

(a)

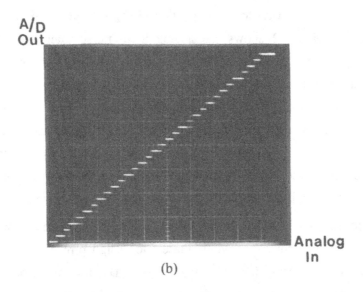

(b)

Fig. 2.9 Measured transfer characteristics of 5-bit neural-based
          A/D converter.
          (a) Original Hopfield A/D converter.
          (b) Modified Hopfield A/D converter.

discrete value of −1 V, 0 V, or 1 V in order to be compatible with the amplifier output voltage and the reference voltage. The circuit diagram using the correction logic is shown in Fig. 2.10.

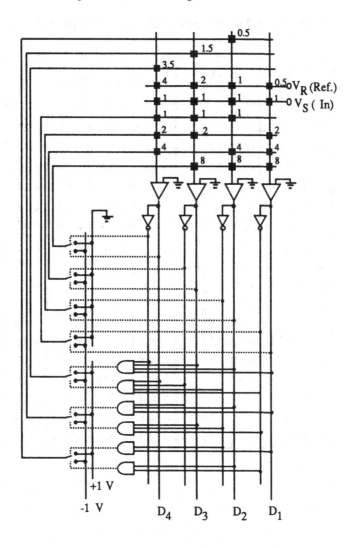

Fig. 2.10 Circuit schematic of a modified 4-bit Hopfield A/D converter with the correction logic gate.

Table 2.2 Analog range after generalized correction

DIGITAL CODE				CORRECTION CURRENT			ANALOG INPUT RANGE			
D_4	D_3	D_2	D_1	I_2^C	I_3^C	I_4^C	V_1	V_2	V_3	V_4
0	0	0	0	+0.5	+1.5	+3.5	< 0.5	< 0.5	< 0.5	< 0.5
0	0	0	1	+0.5	+1.5	+3.5	> 0.5	< 1.5	< 1.5	< 1.5
0	0	1	0	-0.5	+1.5	+3.5	< 2.5	> 1.5	< 2.5	< 2.5
0	0	1	1	+0.5	+1.5	+3.5	> 2.5	> 2.5	< 3.5	< 3.5
0	1	0	0	-0.5	-1.5	+3.5	< 4.5	< 4.5	> 3.5	< 3.5
0	1	0	1	+0.5	-1.5	+3.5	> 4.5	< 5.5	> 4.5	< 5.5
0	1	1	0	-0.5	-1.5	+3.5	< 6.5	> 5.5	> 5.5	< 6.5
0	1	1	1	+0.5	-1.5	+3.5	> 6.5	> 6.5	> 6.5	< 7.5
1	0	0	0	-0.5	+1.5	-3.5	< 8.5	< 8.5	< 8.5	> 7.5
1	0	0	1	+0.5	+1.5	-3.5	> 8.5	< 9.5	< 9.5	> 8.5
1	0	1	0	-0.5	+1.5	-3.5	< 10.5	> 9.5	< 10.5	> 9.5
1	0	1	1	+0.5	+1.5	-3.5	> 10.5	> 10.5	< 11.5	> 10.5
1	1	0	0	-0.5	-1.5	-3.5	< 12.5	< 12.5	> 11.5	> 11.5
1	1	0	1	+0.5	-1.5	-3.5	> 12.5	< 13.5	> 12.5	> 12.5
1	1	1	0	-0.5	-1.5	-3.5	< 14.5	> 13.5	> 13.5	> 13.5
1	1	1	1	-0.5	-1.5	-3.5	> 14.5	> 14.5	> 14.5	> 14.5

Table 2.3 Logic table for the correction logic circuitry

AMPLIFIER OUTPUT	CORRECTION LOGIC OUTPUT		
$D_4\ D_3\ D_2\ D_1$	C_4	C_3	C_2
0 0 0 0	0	0	0
0 0 0 1	0	0	+1
0 0 1 0	0	0	-1
0 0 1 1	0	+1	0
0 1 0 0	0	-1	0
0 1 0 1	0	0	+1
0 1 1 0	0	0	-1
0 1 1 1	+1	0	0
1 0 0 0	-1	0	0
1 0 0 1	0	0	+1
1 0 1 0	0	0	-1
1 0 1 1	0	+1	0
1 1 0 0	0	-1	0
1 1 0 1	0	0	+1
1 1 1 0	0	0	-1
1 1 1 1	0	0	0

To perform SPICE circuit simulation [9-12], amplifiers were modeled as dependent voltage sources, and the rest of the circuit was described at the transistor level. In our simulation, normalized

conductances were used. Figure 2.11 shows the simulated voltage
transfer characteristics of the neural-based A/D converter. In this case, a
monotonically increasing and decreasing analog input voltage was
applied. The simulation results for the Hopfield A/D converter are plotted in solid lines, while those for the modified A/D converter are plotted
in dotted lines. The SPICE results confirm the hysteresis and nonlinearity characteristics of the Hopfield A/D converter. Simulation results of
the modified A/D converter show good output characteristics.

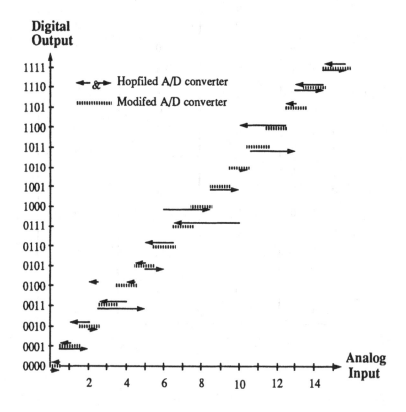

Fig. 2.11 SPICE simulation results on the transfer characteristics of
the original Hopfield A/D converter and the modified A/D
converter.

Figure 2.12 shows a die photo of the modified A/D converter. The chip size is 2300 μm x 3400 μm in MOSIS 3-μm CMOS process [13]. The electronic synapses are realized with p-well diffusion resistors and the electronic neurons are implemented with simple two-stage amplifiers.

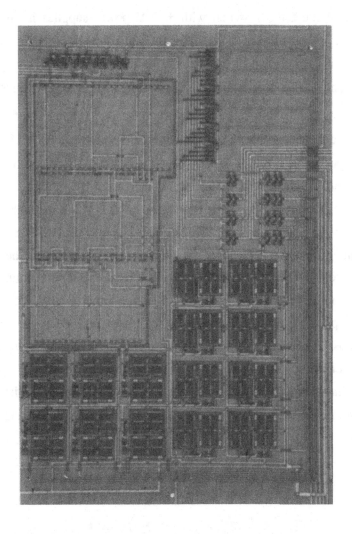

Fig. 2.12 Die photo of the modified A/D converter with
correction logic gate.

Since a simple CMOS amplifier has a large output impedance, it cannot directly drive the feedback conductances. Big CMOS switches between the amplifiers and the resistors help to achieve good impedance matching. The CMOS switch size and conductance values are design parameters. In this chip, the unit resistor is chosen to be 100 kΩ and the on-resistance of the CMOS switch with ± 5 V power supplies is chosen to be 500 Ω. A significant portion of the chip area is occupied by the resistors and switches. If the resistor is replaced by the synthesized resistors using MOS transistors, the chip size can be greatly reduced.

The measured voltage transfer curves for the original Hopfield A/D converter and the modified A/D converter are shown in Fig. 2.13. The analog input voltage range in this experiment was from 0 V to 1.5 V, because the conductances $\{T_{is}\}$ were increased by a factor of ten. Hence, the conversion step size is reduced to 0.1 V. Experimental data agree very well with theoretically calculated results and SPICE simulation results. Figure 2.14 shows the converter response when the output changes from (0000) to (0111) and vice versa. The maximum delay time is about 5.7 μsec and total power dissipation is 6 mW with ± 5 V power supplies. Since the amplifier is the major time delay element, the modified A/D converter has a similar conversion speed. Small spikes on the transfer curve are caused by device mismatches from the amplifier offset voltage, amplifier gain, and resistance. The effect of mismatches can be eliminated by the learning process if programmable synapses are available.

2.5 Traveling Salesman Problem

The traveling salesman problem (TSP) is a well-known np-complete problem. The objective of the TSP is to find the shortest round-trip distance visiting all cities once. However, only an acceptable solution of

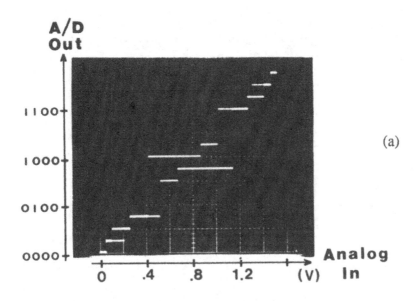

(a)

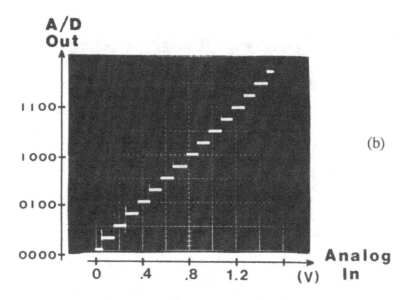

(b)

Fig. 2.13 Measured transfer characteristics of neural-based
A/D converter.
(a) Original Hopfield network.
(b) Modified Hopfield network.

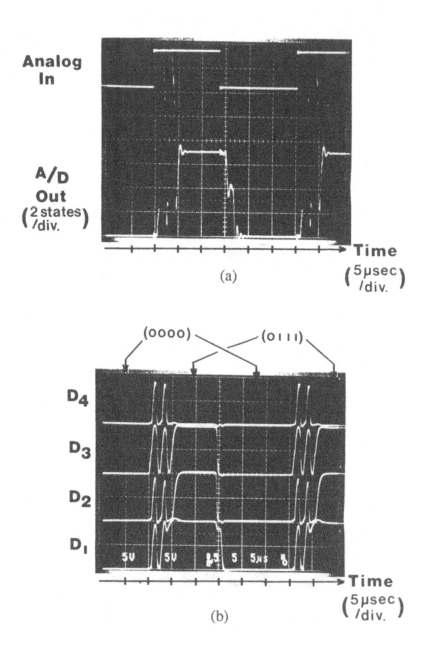

Fig. 2.14 Transient response.
          (a) A/D converter output.
          (b) Individual amplifier outputs.

the TSP with a large city number can be obtained at a reasonable computational time. The Hopfield network can be used to efficiently solve the TSP [14]. For n-city problem, n^2 neurons (n neurons per city) are utilized to represent tour sequence so that the final solution can be shown in the $n \times n$ square array format. The energy function is expressed as

$$E = \frac{A}{2} \sum_{X=1}^{n} \sum_{i=1}^{n} \sum_{j \neq i, j=1}^{n} V_{Xi} V_{Xj} + \frac{B}{2} \sum_{i=1}^{n} \sum_{X=1}^{n} \sum_{Y \neq X, Y=1}^{n} V_{Xi} V_{Yi}$$

$$+ \frac{C}{2} \left[\sum_{X=1}^{n} \sum_{i=1}^{n} V_{Xi} - n \right]^2$$

$$+ \frac{D}{2} \sum_{X=1}^{n} \sum_{Y \neq X, Y=1}^{n} d_{XY} V_{Xi} (V_{Y,i+1} + V_{Y,i-1}) , \qquad (2.52)$$

where X and Y represent cities, i and j represent tour procedure, and d_{XY} is the distance between city X and Y. The first three terms in (2.52) are constraint functions from only one-visit-at-one-city, only one-city-at-one-time, and total visiting number, respectively. Notice that the first three terms become zero when the solution is valid. On the other hand, the last term is the total tour distance which should be minimized. From the energy function, the conductances and the input currents of the Hopfield network in (2.1) are given as

$$T_{Xi,Yj} = - A \delta_{XY} (1-\delta_{ij}) - B \delta_{ij} (1-\delta_{XY}) - C$$

$$- D d_{XY} (\delta_{j,i+1} + \delta_{j,i-1}) \qquad (2.53)$$

and

$$I_{Xi} = Cn . \qquad (2.54)$$

Here, δ_{ij} is 1 if $i=j$ and is 0 otherwise. Since every $T_{Xi,Yj} < 0$, the Hopfield network has local minima from (2.10).

The heuristic coefficients A, B, C, and D should be carefully selected to obtain a valid tour. But, the choice of coefficients becomes very difficult as the number of cities increases [15]. This is caused by a mixture of the constraint functions and the objective function in the energy function. During the search process to find a low energy well, the network cannot distinguish between the constraint function and the objective function. Thus, the final solution is usually invalid when the number of cities is large. Several research results [16,17] have been reported to find a valid solution.

2.5.1 Competitive-Hopfield Network Approach

A new solution consisting of the utilization of a Hopfield network and a competitive network has been developed [18]. The constraint functions are implemented in the competitive network, while the energy function with reduced constraint functions is implemented in the original Hopfield network, as shown in Fig. 2.15. The competitive network monitors the outputs of the Hopfield network. Once the outputs are larger than a threshold voltage of neurons in the competitive network, the network starts to search for the highest output through the competing process with the 'winner-take-all' strategy. Here, signal delay in the competitive network is much smaller than that in the Hopfield network. Since the constraint functions of a combinatorial optimization problem are realized by the competitive network, the energy function of Hopfield network can be minimized. Hence, a solution searching process in the Hopfield network can be determined mainly by the objective function.

For an n-city TSP, the synapses $\{T^H_{Xi,Yj}\}$ and input currents $\{I_{Xi}\}$ for the Hopfield network are given as

$$T^H_{Xi,Yj} = -A\,\delta_{XY}(1-\delta_{ij}) - B\,\delta_{ij}(1-\delta_{XY}) - C$$

$$- Dd_{XY}(\delta_{j,i+1}+\delta_{j,i-1}) \tag{2.55}$$

and

$$I_{Xi} = Cn + (A + B + C + D)V^C_{Xi}, \tag{2.56}$$

while the synapses $\{T^C_{Xi,Yj}\}$ for the competitive network are given as

$$T^C_{Xi,Yj} = \delta_{XY}(1 - \delta_{ij}) + \delta_{ij}(1 - \delta_{XY}) . \tag{2.57}$$

Here, V^C_{Xi} is the output of the competitive network and δ_{ij} is 1 if $i=j$ and is 0 otherwise. Notice that the weighting factors A, B, C, and D are the same as in (2.53).

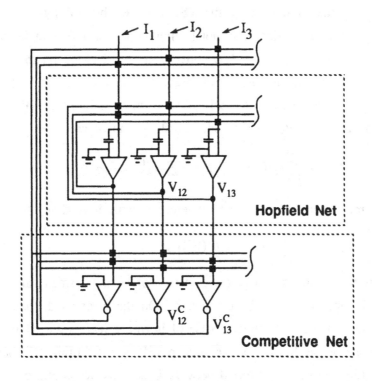

Fig. 2.15 A competitive-Hopfield neural network

Dynamics of the Hopfield network can be described by the following differential equation,

$$C_{Xi} \frac{du_{Xi}}{dt} = \sum_{Y \neq X, Y=1}^{n} \sum_{j \neq i, j=1}^{n} T_{Xi, Yj}^{H} V_{Yj} - T_{Xi} u_{Xi} + I_{Xi} , \qquad (2.58)$$

where T_{Xi}, u_{Xi}, and C_{Xi} are the equivalent conductance, input voltage, and input capacitance at the i-th amplifier input node, respectively. To maintain the Hopfield network dynamics, the competitive network operates after each Hopfield network iteration in the software simulation.

2.5.2 Search for Optimal Solutions

The global minimum of the TSP can be obtained by adding the correction currents. Since the global minimum of a TSP is unknown in contrast to the neural-based A/D conversion, the decision on the correction current is not obvious. However, the decision can be done by adjusting the amount of the correction currents as follows [19]. The lower bound of GAP_k for the codes in (2.13) can be given as

$$GAP_k = \sum_{j=1}^{k-1} T_{kj} (V_j^u - V_j^l) \geq -\sum_{j=1}^{k-1} |T_{kj}^H| \equiv \hat{T}_{kC} . \qquad (2.59)$$

The correcting current I_{kC} can be determined with a weighting factor α,

$$I_{kC} = \alpha \hat{T}_{kC} \left[f^u(V_k) - f^l(V_k) \right] . \qquad (2.60)$$

Here, $\alpha \leq 1.0$. In the undercorrected case of having a small value of α, all local minima can not be eliminated. On the other hand, there is no stable state in the overcorrected case of having a large value of α. The appropriate value for α can be determined by monitoring whether the energy function of the corrected network is steadily decreasing or not.

Experimental results for 10-city and 20-city are shown in Figs. 2.16 - 2.18. The first city on the tour was fixed and the competitive algorithm was applied [18]. For the 10-city TSP, a global minimum can be obtained with a certain weighting factor for the correction currents. On the other hand, the global minimum of the 20-city TSP is not obtained, but significant improvement in the tour distance is made with the correcting currents. When the weighting factor α is very large, the solution becomes poor because the overcorrection can change the global minimum.

Since the lower bound of GAP_k in (2.59) increases with k, the correction of the neuron with high k is usually much larger than that of the neuron with low k. Therefore, the convergence speed to the final value decreases with the k value. A new statistically averaged value for the GAP calculation can be used for the correction current as following,

$$I_{kC} = \alpha \frac{\hat{T}_{kC}}{k} \left[f^u(V_k) - f^l(V_k) \right] . \tag{2.61}$$

Here, $\bar{\alpha}$ can be larger than 1.0. The simulation results using the averaged correction scheme are shown in Figs. 2.19 and 2.20. The minimum tour distance of the averaged scheme is usually smaller than that of the absolute scheme in (2.60).

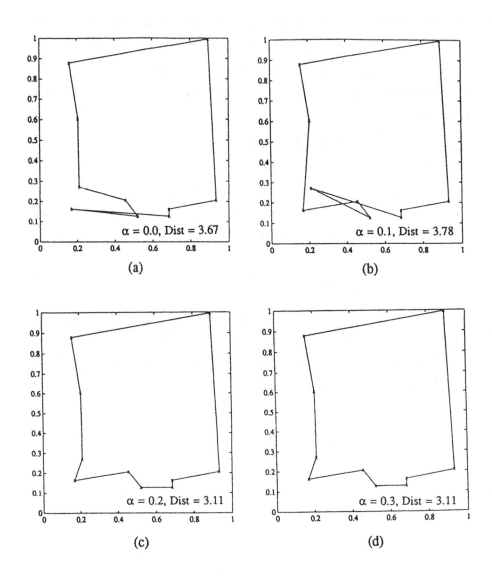

Fig. 2.16 One 10-city TSP.
With $\alpha = 0.2$ and 0.3, the global minimum is obtained.

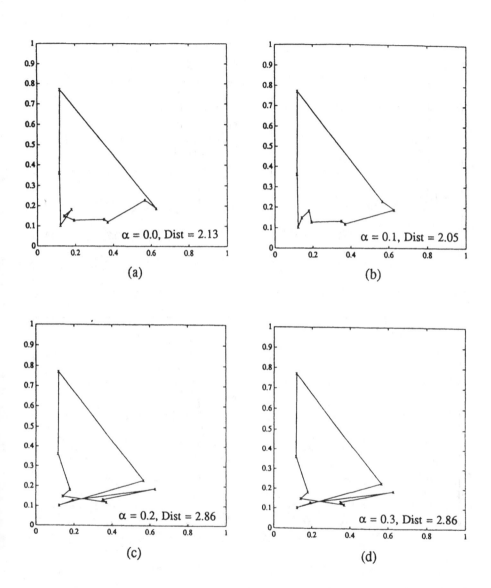

Fig. 2.17 Another 10-city TSP.

With $\alpha = 0.1$, the global minimum is obtained.

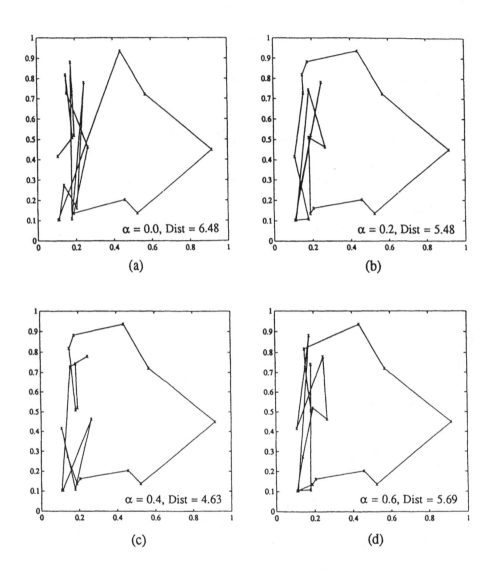

Fig. 2.18 A 20-city TSP.
With $\alpha = 0.4$, a very good solution is obtained.

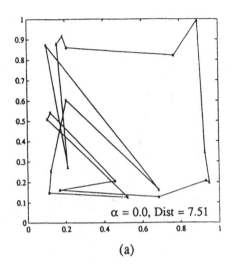

(a)

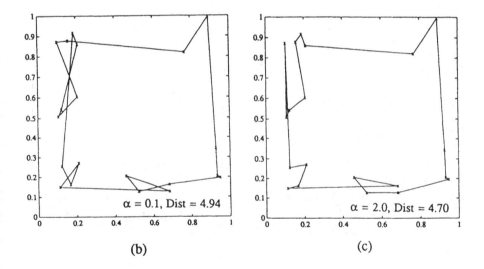

(b)   (c)

Fig. 2.19 One 20-city TSP with averaged correction scheme.

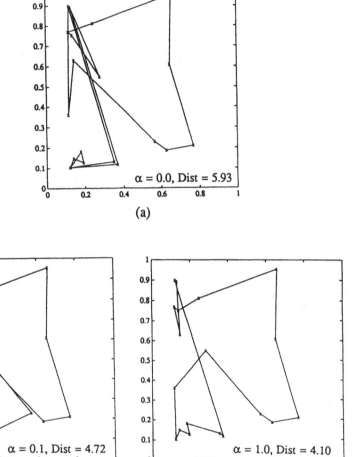

Fig. 2.20 Another 20-city TSP with averaged correction scheme.

References

[1] J. J. Hopfield, "Neurons with graded response have collective computational properties like those of two-state neurons," *Proc. Natl. Acad., Sci. U.S.A.,* vol. 81, pp. 3088 -3092, May 1984.

[2] R. P. Lippman, "An introduction to computing with neural nets," *IEEE Acoustics, Speech, and Signal Processing Magazine,* pp. 4-22, April 1987.

[3] D. W. Tank and J. J. Hopfield, "Simple 'neural' optimization networks: an A/D converter, signal decision circuit, and a linear programming circuit," *IEEE Trans. on Circuits and Systems,* vol. CAS-33, no. 5, pp. 533-541, May 1986.

[4] B. W. Lee, J.-C. Lee, and B. J. Sheu, "VLSI image processors using analog programmable synapses and neurons," *Proc. of IEEE/INNS Inter. Conf. on Neural Networks,* vol. II, pp. 575-580, San Diego, CA, June 1990.

[5] W.-C. Fang, B. J. Sheu, and J.-C. Lee, "Real-time computing of optical flow using adaptive VLSI neuroprocessors," *Proc. of IEEE Inter. Conf. on Computer Design'* Cambridge, MA, Sept. 1990.

[6] J.-C. Lee and B. J. Sheu, "Analog VLSI neuroprocessors for early vision processing," in *VLSI Signal Processing IV,* Editor: K. Yao, New York: IEEE Press, Nov. 1990.

[7] B. W. Lee and B. J. Sheu, "An investigation on local minima of Hopfield network for optimization circuits," *Proc. of IEEE Inter. Conf. on Neural Networks,* vol. I, pp. 45-51, San Diego, CA, July 1988.

[8]   B. W. Lee and B. J. Sheu, "Design of a neural-based A/D converter using modified Hopfield network," *IEEE Jour. of Solid-State Circuits,* vol. 24, no. 4, pp. 1129-1135, Aug. 1989.

[9]   T. Quarles, SPICE3 Version 3C1 Users Guide, *Electron. Res. Lab. Memo UCB/ERL M89/46,* University of California, Berkeley, Apr. 1989.

[10]   A. Vladimirescu, S. Liu, "The simulation of MOS integrated circuits using SPICE2," *Electron. Res. Lab. Memo ERL-M80/7,* University of California, Berkeley, Oct. 1980.

[11]   B. J. Sheu, D. L. Scharfetter, P. K. Ko, M.-C. Jeng, "BSIM: Berkeley short-channel IGFET model for MOS transistors," *IEEE Jour. of Solid-State Circuits,* vol. SC-22, no. 4, pp. 558-566, Aug. 1987.

[12]   HSPICE Users' Manual H9001, Meta-Software Inc., Campbell, CA, May 1989.

[13]   C. Tomovich, "MOSIS-A gateway to silicon," *IEEE Circuits and Devices Magazine,* vol. 4, no. 2, pp. 22-23, Mar. 1988.

[14]   J. J. Hopfield and D. W. Tank, " 'Neural' computation of decisions in optimization problems," *Biol. Cybernetics,* vol. 52, pp. 141-152, 1985.

[15]   G. V. Wilson and G. S. Pawley, "On the stability of the traveling salesman problem algorithm of Hopfield and Tank," *Biol. Cybernetics,* vol. 58, pp. 63-70, 1988.

[16]   D. E. Van den Bout and T. K. Miller, "A traveling salesman objective function that works," *Proc. of IEEE Inter. Conf. on Neural Networks,* vol. II, pp. 299-303, San Diego, CA, July 1988.

[17] Y. Akiyama, A. Yamashita, M. Kajiura, and H. Aiso, "Combinatorial optimization with Gaussian Machines," *Proc. of IEEE/INNS Inter. Joint Conf. on Neural Networks,* vol. I, pp. 533-540, Washington D.C., June 1989.

[18] B. W. Lee and B. J. Sheu, "Combinatorial optimization using competitive-Hopfield neural network," *Proc. of IEEE/INNS Inter. Joint Conf. on Neural Networks,* vol. II, pp. 627-630, Washington D.C., Jan. 1990.

[19] S. M. Gowda, B. W. Lee, and B. J. Sheu, "An improved neural network approach to the traveling salesman problem," *Proc. of IEEE Tencon,* pp. 552-555, Bombay, India, Nov., 1989.

Chapter 3

Hardware Annealing Theory

Engineering optimization is an important subject in signal and image processing. A conventional searching technique for finding the optimal solution is to use gradient descent, which finds a direction for the next iteration from the gradient of the objective function. For complicated problems, the gradient descent technique often gets stuck at a local minimum where the objective function has surrounding barriers. In addition, the complexity of most combinational optimization problems increases dramatically with the problem size and makes it very difficult to obtain the global minimum within a reasonable amount of computational time. Several methods have been reported to assist the network output to escape from the local minima [1,2]. For example, the simulated annealing method is a heuristic approach which can be widely applied to the combinational optimization problems [3,4]; the solutions by the simulated annealing technique are close to the global minimum within a polynomial upper bound for the computational time and are independent of initial conditions; and the simulated annealing technique has been successfully applied in VLSI layout generation [5] and noise filtering in image processing [6].

3.1 Simulated Annealing in Software Computation

Simulated annealing in software computation of neural networks can be conducted in the following way:

Step 1. Start from a high temperature and a given reference state.

Step 2. Compare the energy value in a new state with that in the reference state.

Step 3. If energy value of the new state is higher than that of the reference state, then, weight the new state output by $e^{-\frac{\Delta E}{kT}}$.

Step 4. Replace the reference state with the new state.

Step 5. Check whether all states are frozen.

　　　　If yes, terminate.

　　　　Otherwise, decrease temperature and go to "Step 2."

Since it usually takes a lot of time at "Step 2" to compare all possible states, a specific perturbation rule is often used. The perturbation is usually called artificial noise in software computation.

The simulated annealing technique can help Hopfield neural networks to escape from local minima by replacing the transfer characteristic of neurons from a sigmoid function [7] to the Boltzmann distribution function. In software computation, the Hopfield network operation is described at two consecutive time steps during each iteration cycle. At the first time step, input signals to the neurons are summed up, while at the second step the neuron outputs are updated. The update rule for the original Hopfield networks is

$$V_i = g(u_i) \ , \tag{3.1}$$

and that for the Boltzmann machine is

$$V_i = \frac{1}{1 + e^{-u_i q/kT}} \ . \tag{3.2}$$

Here, $g(.)$ is the amplifier input-output transfer function, q is electronic charge, k is Boltzmann constant, and T is the temperature of Boltzmann distribution as shown in Fig. 3.1. Notice that the searching process for all possible states in simulated annealing is a part of network operation at a given temperature in the Boltzmann machine. At the steady state, the relative probability of state $S1$ to state $S2$ in Boltzmann distribution is determined by the energy difference of these two states,

$$\frac{P_{S1}}{P_{S2}} = e^{-(E_{S1} - E_{S2})/kT} .$$

(3.3)

Here, E_{S1} and E_{S2} are the corresponding energy levels. This update rule allows the network to escape from local minima in the energy well.

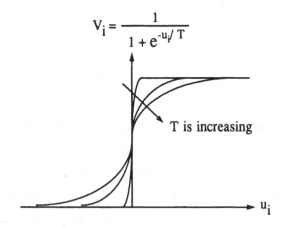

Fig. 3.1   Analogy between the annealing temperature of
a Boltzmann machine and the amplifier gain of
an electronic neuron.

Simulated annealing is analogous to metallurgical annealing. To find the lowest energy level of a metal, the best way known is to melt the metal and to reduce the temperature slowly in order to allow atoms to fit into the lattice. Similarly, the Boltzmann machine can find the global minimum by changing the temperature of the update rule gradually. A good strategy to apply the simulated annealing technique in software computation is to start from a high temperature, which will make the network reach the steady state in a very short time. An ideal formulation for the cooling schedule [3] is very difficult to implement, because a large number of iterations at each temperature and a very small temperature step are required to achieve the global minimum. Several approximated cooling schedules have been proposed [2]. Since the number of iterations at a given temperature and the change of temperature should be compromised to speed up the convergence process, a very large computational time is necessary in software computation. However, recent advances in VLSI technologies make possible the design of compact electronic neural networks with built-in hardware annealing capability.

3.2 Hardware Annealing

The high-speed simulated annealing technique for a Boltzmann machine is most suitable for VLSI chip design [8]. Changing the temperature of the probability function for a Boltzmann machine is equivalent to varying the amplifier gain. Thus, the cooling process in a Boltzmann machine is equivalent to the voltage gain increase process in an amplifier. The amplifier gain in electronic neural circuits can be adjusted continuously, while the annealing temperatures in software computation are always adjusted in a discrete fashion. Hence, the final solution of the electronic neural circuits after annealing is guaranteed to be a global minimum, which is in sharp contrast to the approximated convergence in software computation.

A systematic method to determine the initial and final temperatures for the amplifier "cooling" procedure is described below.

3.2.1 Starting Voltage Gain of the Cooling Schedule

At a very high temperature, all metal atoms lose the solid phase so that they position themselves randomly according to statistical mechanics. An important quantity in annealing is the lowest temperature that can provide enough energy to completely randomize the metal atoms; equivalently, the highest amplifier gain to make an electronic neural network escape from local minima. Figure 3.2 shows a Hopfield neural network. Neurons are made of amplifiers while synapses are made of resistors. If the neurons consist of high-gain amplifiers, their outputs will saturate at 0 V and 1 V due to the positive feedback of the network. However, given a very low amplifier gain, the network could lose the well-defined state.

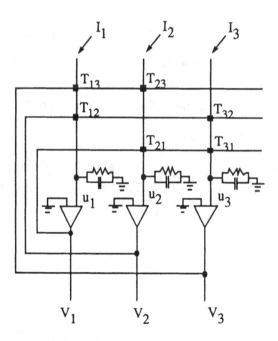

Fig 3.2   A Hopfield neural network with non-inverting
amplifiers as neurons.

Assuming that all amplifiers operate in the high-gain region, the governing equation for the i-th amplifier in the Hopfield network is given as

$$C_i \frac{du_i(t)}{dt} + T_i u_i(t) = \sum_{j \neq i, \, j=1}^{n} T_{ij} v_j(t) + I_i(t) , \qquad (3.4)$$

where T_{ij} is the conductance between the i-th and j-th neurons and T_i and C_i are the equivalent input conductance and input capacitance to the i-th amplifier. Here, $u_i(t)$ is the input voltage to the i-th amplifier and $v_j(t)$ is the output voltage of the j-th amplifier. Notice that $v_j(t)$ denotes the analog value and $V_j(t)$ denotes the saturated digital value. By taking Laplace transformation, (3.4) becomes

$$(sC_i + T_i)U_i(s) = \sum_{j \neq i, \, j=1}^{n} T_{ij} V_j(s) + I_i(s) + P_i , \qquad (3.5)$$

where P_i is a constant and $U_i(s)$, $v_i(s)$, and $I_i(s)$ are transformed variables of $u_i(t)$, $V_i(t)$, and $I_i(t)$, respectively. If all amplifiers are assumed to operate in the high-gain region with the transfer function being $A(s)$ and to have bandwidth much larger than T_i/C_i, then

$$V_i(s) = A_i(s)U_i(s) . \qquad (3.6)$$

The system equation can be expressed as $\mathbf{B} \underline{V} = \underline{F}$ with matrix $\mathbf{B}$ being an $n \times n$ matrix and $\underline{V}$ and $\underline{F}$ being $1 \times n$ vectors. The matrix and vector expressions for $\mathbf{B}$, $\underline{V}$, and $\underline{F}$ are

$$\mathbf{B} = \begin{bmatrix} -\dfrac{sC_1 + T_1}{A_1} & T_{12} & T_{13} & \cdot & \cdot & T_{1n} \\[2ex] T_{21} & -\dfrac{sC_2 + T_2}{A_2} & T_{23} & \cdot & \cdot & T_{2n} \\[2ex] \cdot & \cdot & \cdot & \cdot & \cdot & \cdot \\ \cdot & \cdot & \cdot & \cdot & \cdot & \cdot \\ \cdot & \cdot & \cdot & \cdot & \cdot & \cdot \\ T_{n1} & T_{n2} & \cdot & \cdot & \cdot & -\dfrac{sC_n + T_n}{A_n} \end{bmatrix} \quad (3.7)$$

$$\underline{V} = \begin{bmatrix} V_1, V_2, \ . \ . \ . \ , V_n \end{bmatrix}^T , \tag{3.8}$$

and

$$\underline{F} = \begin{bmatrix} -I_1 - P_1, \ . \ . \ . \ . \ , -I_n - P_n \end{bmatrix}^T . \tag{3.9}$$

The sum and product of eigenvalues of the system matrix B are

$$\sum_{i=1}^{n} \lambda_i = -\sum_{i=1}^{n} \frac{(sC_i + T_i)}{A_i} \tag{3.10}$$

and

$$\prod_{i=1}^{n} \lambda_i = \det(\mathbf{B}) \tag{3.11}$$

respectively, where $\det(\mathbf{B})$ is the determinant of matrix $\mathbf{B}$. If the amplifier gain is sufficiently large, which is the condition used in Hopfield's analysis [9], (3.10) and (3.11) become

$$\sum_{i=1}^{n} \lambda_i \approx 0 \tag{3.12}$$

and

$$\prod_{i=1}^{n} \lambda_i = \det(\mathbf{B}) \neq 0. \tag{3.13}$$

With the constraint that $T_{ij} = T_{ji}$, all eigenvalues will lie on the real axis of the s-plane. Thus, at least one positive real eigenvalue exists. It makes the amplifier outputs saturated at extreme values of 0 V or -1 V.

Figure 3.3 shows the radius of the eigenvalues determined from the Gerschgorin's Theorem [10],

$$ \mid z + (sC_i + T_i) \mathbin{/} A_i \mid \leq \sum_{j \neq i,\, j=1}^{n} \mid T_{ij} \mid \quad \text{for } all \; i \; . \qquad (3.14) $$

Notice that A_i and T_i are always positive. To assure that the network operates in the positive feedback mode, there should be at least one eigenvalue whose real part is positive. The lowest amplifier gain (A_n) which satisfies the above condition can be determined from

$$ A_n = \max \left\{ \frac{T_i}{\displaystyle\sum_{j \neq i,\, j=1}^{n} \mid T_{ij} \mid} \quad \text{for } 1 \leq i \leq n \right\} . \qquad (3.15) $$

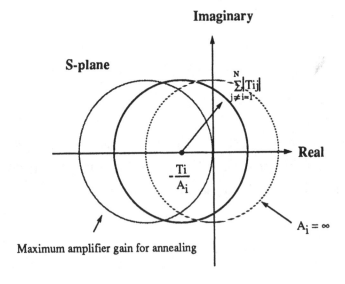

Fig. 3.3 Radius of eigenvalues with different amplifier gains.

Here, n denotes the number of amplifiers operating in the high-gain region. Notice that the above derivation is based on the condition that the eigenvalue lies on a circle. Since validity of the condition depends on the maximum real value of the eigenvalues of matrix $\mathbf{B}$ determined by the resistive network $\{T_{ij}\}$, the above gain requirement is a sufficient condition that the amplifiers in Hopfield network stay in the positive feedback mode. With the amplifier gain less than A_n, all output states of the Hopfield network become legal for any input signal value. Thus, A_n is the maximum amplifier gain (equivalently the lowest annealing temperature) that can still randomize the neuron outputs.

3.2.2 Final Voltage Gain of the Cooling Schedule

The amplifiers start to be biased in the saturation region for a given input voltage if amplifier gains are increased from A_n. Let's assume that only the k-th amplifier output is saturated at a digital value V_k. The governing equations for the new network condition are

$$C_i \frac{du_i(t)}{dt} + T_i u_i(t) = \sum_{j \neq i, k; j=1}^{n} T_{ij} v_j(t) + I_i(t) + T_{ik} V_k$$

$$\text{for any } i \neq k. \tag{3.16}$$

The corresponding system matrix can be formed with the k-th column and the k-th row being deleted from (3.7). Therefore, the lowest amplifier gain which makes the network stay in the positive feedback mode is determined by

$$A_{n-1} = \max \left\{ \frac{T_i}{\displaystyle\sum_{j \neq i, k; j=1}^{n} |T_{ij}|} \quad \text{for } 1 \leq i \leq n \right\}. \tag{3.17}$$

The same T_i is used in (3.15) and (3.17). Notice that the amplifier gain for the positive feedback state increases as the number of linear region amplifiers decreases. The worst case occurs when only two amplifiers operate in the high-gain region. Let's assume that the i-th and k-th amplifiers operate in the high-gain region. The system matrix $\mathbf{B_{ik}}$ can be expressed as

$$\mathbf{B_{ik}} = \begin{bmatrix} -\dfrac{sC_i + T_i}{A_i} & , & T_{ik} \\[2ex] T_{ki} & & -\dfrac{sC_k + T_k}{A_k} \end{bmatrix}. \tag{3.18}$$

Since the resistive network $\{\ T_{ij}\ \}$ is symmetrical, $T_{ik} \times T_{ki}$ is always positive. The maximum gain for two high-gain region amplifiers is

$$A_2 = \max\left\{ \sqrt{\frac{T_i T_k}{T_{ik} T_{ki}}} \right\} \quad \text{for every } i \text{ and } k \text{ with } i \neq k \ . \tag{3.19}$$

When the amplifier gain is increased beyond A_2, only one amplifier will operate in the high-gain region. Even though the amplifier output is an analog value, the digital bit can be easily decided using the middle level of amplifier output swing as a reference. The logical state of the amplifier biased in the high-gain region can then be determined.

The amplifier gain during the hardware annealing process should start from a value smaller than A_n and stop at a value larger than A_2. In a similar way, the annealing temperature range for the Boltzmann machine using the update rule of (3.2) is

$$\frac{1}{4 A_2} \leq T \leq \frac{1}{4 A_n} \ . \tag{3.20}$$

Since the updating process in the hardware annealing can be done in a continuous fashion, its operation speed is much faster than that for the simulated annealing.

3.3. Application to the Neural-Based A/D Converter

The neural-based A/D converter is a good example to examine the effectiveness of hardware annealing because the global minimum at any analog input voltage is well-defined. In addition, the synapse matrix is very stiff, as shown in Fig. 3.4. Due to this stiffness, the convergence at a given temperature takes a very long computational time in the software computation. For simulated annealing, the temperature takes discrete values and changes by a small amount. Thus, the total software computation time is extremely long. On the contrary, the updating of the hardware annealing process can be done continuously as inherent to the network operation. The computational time for hardware annealing is usually many orders of magnitude smaller than that for the software exercise.

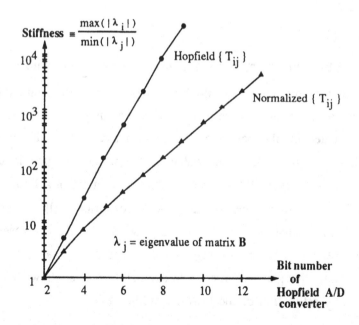

Fig. 3.4 Stiffness of the system function of a neural-based A/D converter.

3.3.1 Neuron Gain Requirement

The maximum neuron gain is determined when only two amplifiers operate in the high-gain region and the others are saturated at logical values. With the maximum gain, one neuron output is saturated, while the other neuron can still operate in a high-gain region. However, the neuron output can be classified into a logical value. Since the synapse weighting $\{T_{ij}\}$ of Hopfield neural-based A/D converter increases with i and j, the maximum gain can be obtained when the i and k are 1 and 2 in (3.19). When the number of bits n is very large, the amplifier gain A_2 is given as

$$A_2 = 2^{n-0.5} . \tag{3.21}$$

For normalized synapse weightings in [9], the maximum neuron gain A_2^{nor} is given as

$$A_2^{nor} = 2^{n-1} . \tag{3.22}$$

Notice that the gain for both cases increases exponentially with n.

Since the formula of the minimum neuron gain to obtain the successful annealing results is derived from the assumption that the eigenvalue is located on the boundary Gerschgorin circle, the value in (3.15) is the lower limit. The actual value of the minimum gain is usually larger than 1.0. Figure 3.5 shows the trajectory of the maximum eigenvalue of the Hopfield neural-based A/D converter [10]. When the amplifier gain decreases below 2, the eigenvalue quickly decreases. On the other hand, the maximum eigenvalue is saturated to a certain value when the amplifier gain is large. The simulation results show that (3.15) gives a good estimation of the lowest amplifier gain.

Figure 3.6 shows the plots of the neural-based A/D converter output as a function of the amplifier gain. In this SPICE circuit simulation, a neuron consists of a simple CMOS operational amplifier with its gain

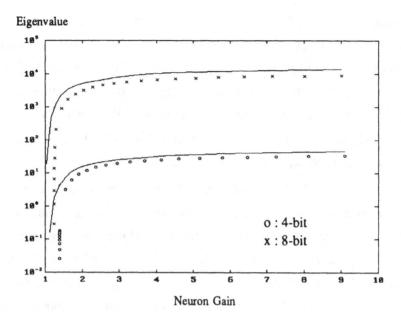

Fig. 3.5 Trajectory of the largest eigenvalue of Hopfield A/D
converter against neuron gain.

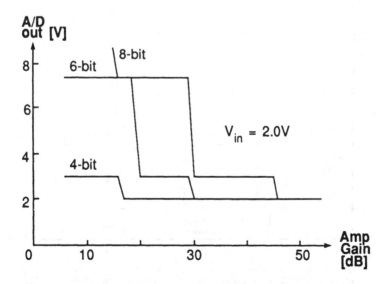

Fig. 3.6 Neural-based A/D converter outputs versus amplifier
gain.

controlled by an MOS transistor at the amplifier input. A triangular voltage waveform is applied to the gate terminal of the controlling transistor to adjust the amplifier gain. Here, the analog input voltage is 2.0 V. For the 8-bit case, the A/D converter output starts to change when the amplifier gain is about 46 dB. In comparison, the theoretically calculated amplifier gain from (3.19) is 45.13 dB. The amplifier gain A_2 increases exponentially with the number of the A/D converter bits.

A 4-bit Hopfield neural-based A/D converter is used to monitor each amplifier behavior. The A/D converter output is reconstructed with a D/A converter. The global minimum is always reached after one annealing cycle, as shown in Fig. 3.7. Here, the initial network state is set to be (0010), which is one of the local minima corresponding to $V_{in} = 1.4\ V$. Figure 3.8 shows the dynamics of the A/D converter with respect to the amplifier gain. As the amplifier gain decreases, the states

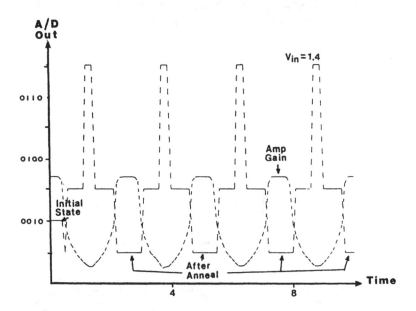

Fig. 3.7 SPICE simulation results of an 4-bit Hopfield neural-based A/D converter in time domain analysis with $V_{in} = 1.4$V.

of amplifier outputs V_1, and V_4 are flipped. The amplifier outputs
remain at these states even when the high voltage gain is restored.

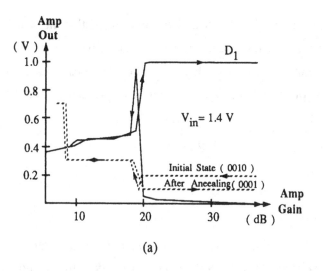

(a)

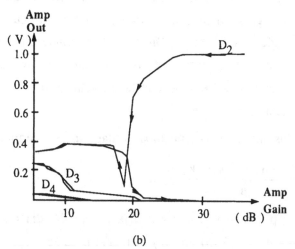

(b)

Fig. 3.8 Trajectories of the A/D converter output versus
the amplifier gain.
(a) Amplifier D_1.
(b) Amplifiers D_2, D_3, and D_4.

3.3.2 Relaxed Gain Requirement Using Modified Synapse Weightings

The large amplifier gain and synapse weighting ratio for a high-bit neural-based A/D converter can be dramatically reduced by using the weighted voltages instead of the weighted synapses. As shown in Fig. 3.9, the neuron output voltages and reference voltages have the following voltage ranges,

$$V_{Oi} = [0 \ V, \ - 2^{i-1} \ V]$$ (3.23)

and

$$V_{Ri} = - 2^{i-2} \ .$$ (3.24)

Here, the voltage sources for the neuron outputs and reference voltages can be generated from a R-2R ladder network. In the direct-R implementation, the large ratio of conductances, which is difficult to obtain in VLSI technologies, was used. On the other hand, only unity-ratioed conductance and binary weighted voltage sources are used in this modified structure. Notice that the large voltage ratio can easily be obtained in VLSI technologies.

Using this structure, neuron input voltage u_i is determined by

$$u_i = \frac{V_{Oi} + V_{Ri} + V_S}{n + 1} \ .$$ (3.25)

Since the synapse matrix is unity matrix except the diagonal terms, the amplifier gains A_n in (3.15) and A_2 in (3.19) become

$$\hat{A}_n = \frac{n + 1}{n}$$ (3.26)

and

$$\hat{A}_2 = n + 1 \ .$$ (3.27)

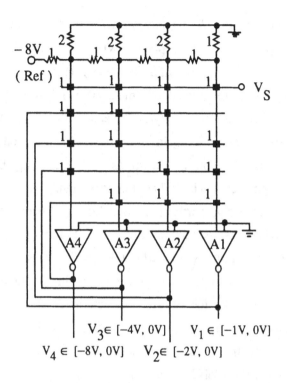

Fig. 3.9 A 4-bit neural-based A/D converter with unity
          synapse weighting.

In the hardware experiment, a 4-bit neural-based A/D converter was constructed using standard integrated circuit parts. The amplifier gain is controlled with a junction-FET operating in the negative feedback loop of the amplifier. Since we used p-channel depletion mode junction-FETs, a high control signal level makes a large amplifier gain. The 4-bit Hopfield neural-based A/D converter outputs were sampled and reconstructed by a D/A converter after every annealing cycle. Transfer characteristics of the A/D converter are shown in Fig. 3.10. The non-linearity and hysteresis of the Hopfield A/D converter is caused by local minima in the energy function. By applying the hardware annealing technique, the global minimum is obtained. Figure 3.11 shows the time domain response of the A/D converter with hardware annealing. Here, two analog input values are applied. When the amplifier gain is high and V_{in} = 3.3 V, the local minimum (1000) is reached. By applying one annealing cycle, the global minimum (0011) is found, as shown in Fig. 3.11(a). Other annealing styles have also been investigated. With abruptly increasing amplifier gain, the output could be either (0011) or (0100). This phenomenon is known as the Quenching Effect in condensed matter physics [11]. When the temperature of the heat bath is lowered instantaneously, particles in a solid are frozen into metastable states. The other case shown in Fig. 3.11(c) and (d) gives the same results. Here, local minimum (0001) at V_{in} = 2.1 V is caused by device mismatches. Even though device mismatches in the neural-based A/D conversion circuit make the transfer characteristics change, the hardware annealing technique helps to find the optimal solution. This hardware experiment demonstrates that only a gradually cooling temperature is effective to reach the global minimum.

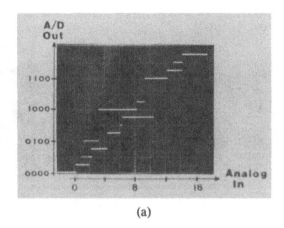

(a)

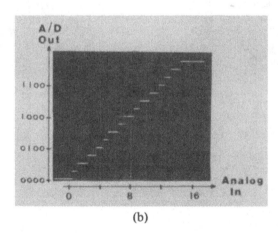

(b)

Fig. 3.10 Transfer characteristics of a 4-bit Hopfield A/D
        converter.
        (a) Without hardware annealing.
        (b) With hardware annealing.

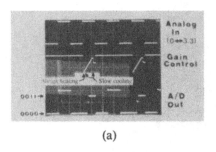

(a)

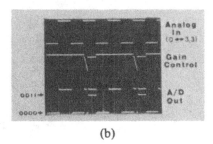

(b)

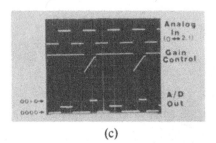

(c)

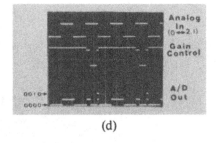

(d)

Fig. 3.11 Different annealing styles.
(a) Fast heating and slow cooling with V_{in} = 0V and 3.3V.
(b) Slow heating and fast cooling with V_{in} = 0V and 3.3V.
(c) Fast heating and slow cooling with V_{in} = 0V and 2.1V.
(d) Fast heating and fast cooling with V_{in} = 0V and 2.1V.

References

[1]   B. W. Lee and B. J. Sheu, "An investigation on local minima of Hopfield network for optimization circuits," *Proc. of IEEE Inter. Conf. on Neural Networks*, vol. I, pp. 45-51, San Diego, CA, July 1988.

[2]   P. J. M. van Laarhoven and E. H. L. Aarts, *Simulated Annealing: Theory and Applications*, Boston, MA: Reidel, 1987.

[3]   S. Kirkpatrick, C. D. Gelatt, Jr., and M. P. Vecchi, "Optimization by simulated annealing," *Science*, vol. 220, no. 4598, pp. 671-680, May 1983.

[4]   E. H. L. Aarts and P. J. M. van Laarhoven, "A new polynomial-time cooling schedule," *Proc. of IEEE Inter. Conf. on Computer-Aided Design*, pp. 206-208, Nov. 1985.

[5]   R. A. Rutenbar, "Simulated Annealing Algorithms: An Overview," *IEEE Circuits and Devices Magazine*, vol. 5, no. 1, pp. 19-26, Jan. 1989.

[6]   S. Geman and D. Geman, "Stochastic relaxation, gibbs distributions, and the bayesian restoration of images," *IEEE Trans. on Pattern Analysis and Machine Intelligence*, vol. PAMI-6, no. 6, pp. 721-741, Nov. 1984.

[7]   D. E. Rumelhart, J. L. McClelland, and the PDP Research Group, *Parallel Distributed Processing*, vol. 1, Cambridge, MA: The MIT Press, pp. 282-317, 1986.

[8]   B. W. Lee and B. J. Sheu, "Hardware simulated annealing in electronic neural networks," *IEEE Trans. on Circuits and Systems*, to appear in 1990.

[9]   J. J. Hopfield, "Neurons with graded response have collective com-
       putational properties like those of two-state neurons," *Proc. Natl.
       Acad., Sci. U.S.A.*, vol. 81, pp. 3088-3092, May 1984.

[10]   G. D. Smith, *Numerical Solution of Partial Differential Equations:
       Finite Difference Methods*, Oxford University Press, pp. 60-63,
       1985.

[11]   D. D. Pollock, *Physical Properties of Materials for Engineers*,
       Boca Raton, FL: CRC Press, pp. 14-18, 1982.

Chapter 4

Programmable Synapses and Gain-Adjustable Neurons

Significant progress has been made in using analog circuit design techniques for early vision processing. The sensory neural networks differ from others in highly localized information processing [1]. At the same time, artificial neural networks for digital signal and image processing usually require heavy interconnections between the neurons. Several experimental neural hardware designs, which typically function as direct accelerators for software computation, have been reported [2-7]. These hardware accelerators consist of main digital processors, mathematical coprocessors, and associated memories. In most cases, the rich properties of neural networks associated with massively parallel processing using programmable synapses and neurons to process digital pictures for machine intelligence have not yet been fully explored.

The endurability for long-term storage and the reprogrammability for adaptive functions of neural networks are necessary properties for VLSI synapses. The amorphous-Si approach from AT&T Bell Labs. has only one-time programming capability, which is similar to the use of read-only memories. In contrast, the charge storage mechanisms used in DRAMs and EEPROMs can be utilized to retain the synapse weighting information. Electronic charge in the DRAM cell is stored at the dynamically refreshed capacitor while the electric power is on. The electronic charge in the EEPROM cell is stored permanently in the floating gate. Some initial investigation results on the design of programmable synapses using the two storage mechanisms have been reported [8-10].

To implement a stochastic neural architecture, a special randomizing circuitry is necessary for the VLSI neural networks. The special circuitry realized in a pseudo-digital format [11] generates a random pulse signal, which is added to the synapse weighting. The random signal generator usually occupies a large silicon area and decreases the neural information processing time. The hardware annealing technique, which needs gain-adjustable neurons, is a simple way to realize the stochastic process. To make the neural computing hardware more powerful, the design and implementation of gain-adjustable neurons and electrically programmable synapses represent a fundamentally important step.

4.1 Compact and Programmable Neural Chips

The 64-neuron VLSI chip from Intel [8] includes a fully programmable synapse with EEPROM cells, as shown in Fig. 4.1. The synapse weighting is coded with the bias currents of the analog multiplier. The input voltage V_i is applied to two differential pairs. Since the EEPROM-injected electronic charge at the floating gate can change the threshold voltage of the transistor in the current mirror circuitry, bias currents of the analog multiplier are programmable. Figure 4.1(b) shows I-V characteristics of the synapse cell for various bias current values.

When V_i is very small, the equivalent conductance T_{eq} can be obtained as

$$T_{eq} \equiv \frac{I_{out}}{V_i} = \sqrt{\mu C_{ox} \frac{W}{L} I_{S1}} - \sqrt{\mu C_{ox} \frac{W}{L} I_{S2}} . \qquad (4.1)$$

Here, I_{S1} and I_{S2} are the programmed currents. The conductance value is determined by the programmed differential current. However, linearity

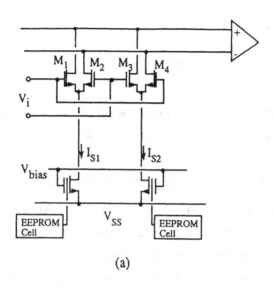

(a)

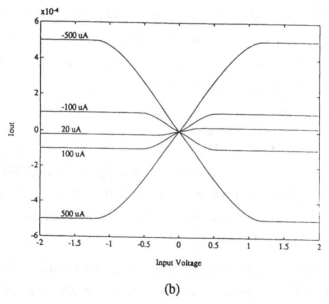

(b)

Fig. 4.1 A programmable synapse cell in [8]
(a) Circuit schematic.
(b) Simulated transfer characteristics.

of the conductance is limited by

$$V_{IR} = \sqrt{\frac{2 \cdot \min(I_{S1}, I_{S2})}{\mu C_{ox} W/L}} . \qquad (4.2)$$

For a proper input voltage range, small device aspect ratio W/L and large programmed currents are required. In addition, the large dynamic range of the conductance value cannot be obtained because the conductance is determined by the square root of the programmed current.

A programmable synapse with very large dynamic range is shown in Fig. 4.2. The transconductance amplifier consisting of $M_1 - M_4$ is connected as a buffer. The information of a synapse weighting is stored at the capacitance associated with the controlling gate in the voltage format. The synapse output current I_{ij}^s of the transconductance amplifier is determined by a function of the differential input voltage and the bias current I_j. When the voltage difference between u_i and V_{ij}^s is small and the logic-1 input is applied, I_{ij}^s is given as

$$I_{ij}^s = \sqrt{\beta I_j} \ (V_{ij}^s - u_i) . \qquad (4.3)$$

Here, β is the transconductance coefficient of M_1 and M_2, u_i is the input voltage of the i-th output neuron, and V_{ij}^s is the programmed voltage for the synapse between the i-th output neuron and j-th input neuron. Voltage V_{ij}^s is stored at the capacitance when selected through the column decoder and the row decoder.

The synthesized synapse cell operates as a three-terminal transconductor. The third terminal is used to change the effective conductance of the programmable synapse cell. For binary input signals, the equivalent model of the programmable synapse is shown in Fig. 4.2(b). Instead of programming actual conductance values, the equivalent current summation can be obtained by using the programming voltage V_{ij}^s with the fixed conductance G_{ij}. Here, G_{ij} is given as

$$G_{ij} = \sqrt{\beta I_j} \; . \tag{4.4}$$

Notice that the current summing function of an artificial neuron can be realized by using either constant input range with programmed synapse weightings or programmed input range with constant synapse weighting.

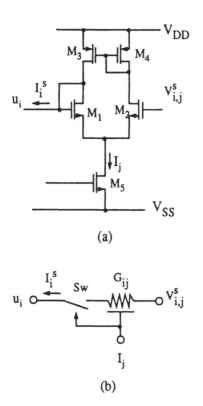

(a)

(b)

Fig. 4.2 Transconductance amplifier as the core for the programmable synapse cell.
(a) Circuit schematic.
(b) Equivalent model for binary inputs.

Figure 4.3 shows the synapse characteristics simulated in SPICE3 and HSPICE circuit simulators [12-15]. The synapse output current is shifted horizontally by the synapse weighting voltage. Since the output neuron voltage is determined by u_i with the threshold at 0 V, the equi-

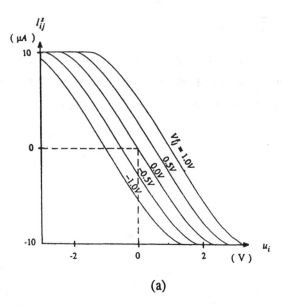

(a)

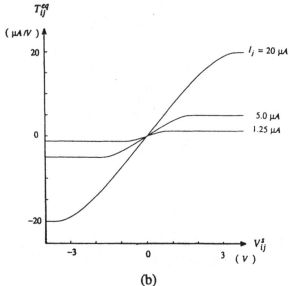

(b)

Fig. 4.3 Simulated synapse characteristics .
    (a) Synapse output current versus neuron input voltage.
    (b) Equivalent synapse conductance versus programmed
        synapse voltage.

valent synapse weighting T_{ij}^{eq} is shown in Fig. 4.3(b).   The dynamic range voltage V_{lmt} of T_{ij}^{eq} is limited by the bias current,

$$V_{lmt} = \sqrt{\frac{2I_j}{\beta}}. \tag{4.5}$$

When u_i is very small, T_{ij}^{eq} is given as

$$T_{ij}^{eq} = \begin{cases} \sqrt{\beta I_j}V_{ij}^s & \text{when } V_j = \text{'logical 1'} \\ 0 & \text{when } V_j = \text{'logical 0'}. \end{cases} \tag{4.6}$$

Notice that the relative value of T_{ij}^{eq} is important in the analog VLSI neural circuits, because u_i is determined by the ratio of synapse weightings.   The maximum value of T_{ij}^{eq} is determined by V_{lmt}, while the minimum value of T_{ij}^{eq} is determined by the resolution of transconductance amplifiers.   Thus, a large dynamic range for the synapse cell can be achieved by using a moderate I^{max} value and a small aspect ratio for transistors M_1 and M_2 in the transconductance amplifier.

The complete circuit schematic including the programmable synapse, input neuron, and gain-adjustable output neuron is shown in Fig. 4.4.   The input neuron consisting of M_8 - M_{10} steers the dynamic range control current I^{max} into the current mirror and produces the bias voltage for multiple synapse cells.   Each synapse output current is summed up and converted to the voltage format at the input terminal of the i-th output neuron.   The output neuron, which is a simple two-stage amplifier with externally adjustable gain, amplifies the converted voltage.

By assuming transconductance amplifiers in synapses $\{T_{ij}; j = 1, 2, ..., n\}$ operating in the high-gain region, the input voltage u_i is governed by the following equation,

$$\left[\sum_{j=1}^n G_{ij}\right] u_i = \sum_{j=1}^n I_{ij}^s . \tag{4.7}$$

Notice that this circuit realizes the synapse weighting in the current format, instead of the direct-R format [16,17]. The minimum amplifier gain for this programmable synapse is determined by the number of synapses. In the case of the direct-R implementation, u_i is given as

$$u_i = \frac{\sum\limits_{j=1}^{n} T_{ij} V_j}{\sum\limits_{j=1}^{n} |T_{ij}|} . \tag{4.8}$$

Since u_i for the direct-R implementation is reduced by the sum of synapse weightings, the minimum amplifier gain for a proper circuit operation is determined by the ratio of the synapse summation and minimum synapse weight [17].

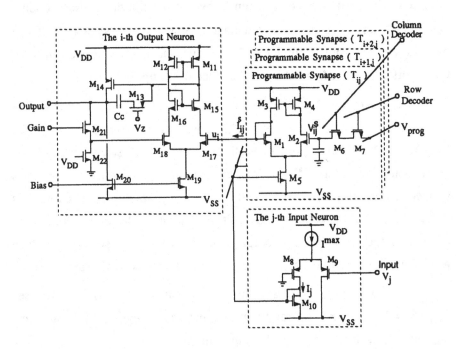

Fig. 4.4 Circuit schematic of the programmable synapse and realted neurons.

Table 4.1. Device geometries of circuit in Fig. 4.4

(unit : μm)

Transistor	W / L	Transistor	W / L
M_1	3/39	M_{12}	6/6
M_2	3/39	M_{13}	3/18
M_3	4/4	M_{14}	6/6
M_4	4/4	M_{15}	6/3
M_5	5/3	M_{16}	6/3
M_6	4/2	M_{17}	18/2
M_7	4/2	M_{18}	18/2
M_8	5/3	M_{19}	6/4
M_9	5/3	M_{20}	6/4
M_{10}	5/3	M_{21}	54/3
M_{11}	6/6	M_{22}	2/80

Capacitor C_c = 60 x 60

The input neuron operates as a buffer which converts input neuron voltage V_j into current I_j. Notice that the actual output voltage range of the neuron amplifier can be [V_{SS}, V_{DD}] while the output current of the input neuron is [0, I^{max}]. For an analog input value of V_j, the input neuron adds a certain voltage gain. The transfer function of the input neuron can be effectively lumped to that of the output neuron in the previous layer of multilayer neural networks. In the Hopfield neural network case, the transfer function of an input neuron is grouped into the output neuron due to the feedback connection. This current driving scheme to connect the input neuron with programmable synapses saves the silicon area tremendously as compared with the dedicated impedance-matching interface [17] between an input neuron and its associated synapses in a direct-R implementation.

The neuron transfer function in software computation is usually specified as an exact mathematical equation, which can decide the neural network properties in the retrieving and learning processes. In our design, transistors M_{15} and M_{16} form an improved cascode stage to increase the voltage gain and M_{13} operates in the triode region to provide the frequency-stabilization resistance. The amplifier voltage gain can be changed by externally applied gain control voltage, which adjusts the drain-to-source resistance ratio of transistors M_{21} and M_{22}. If the gate voltage of M_{21} is below transistor threshold voltage, the amplifier operates in the open-loop high-gain configuration. When the gate voltage of M_{21} increases above the threshold voltage, the closed-loop gain of the amplifier decreases. In the extreme case of having a very large gate voltage applied to M_{21}, the amplifier effectively operates in the unity-gain configuration. The gain-adjustable amplifiers can be used to exercise hardware annealing, which helps the network to efficiently find the global minimum. In our design, the Scalable 2-μm CMOS technology from the MOSIS Service of USC/Information Sciences Institute (Tel. 213-822-1511) was used.

In order to provide a complete analog multiplication capability, a non-saturated version of the programmable synapse cell can be used [18,19]. Figure 4.5 shows the circuit diagram and simulated output characteristics of the non-saturated synapse cell. Additional synapse current is provided by transistors M_5 to M_8 for large differential input voltages.

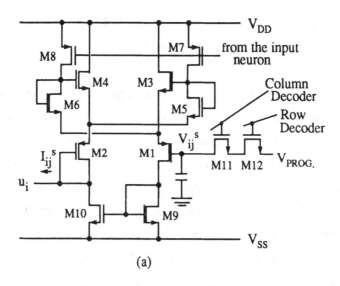

(a)

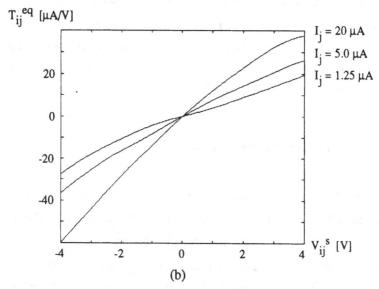

(b)

Fig. 4.5 The programmable synapse cell consisting of a non-
saturated trasnconductance amplifier.
(a) Circuit schematic.
(b) Simulated synapse conductance versus programmed
synapse voltasge.

4.2 Medium-Term and Long-Term Storage
of Synapse Weight

In a biological system, the neuron holds the processing data temporarily through the summing and thresholding steps while the synapse weight keeps a long-term history of network behaviors. Thus, the biological neuron and synapse are called short-term memory and long-term memory, respectively. A similar data storage function can be achieved in the artificial VLSI neural networks. The input capacitance and the finite time delay of a neuron amplifier correspond to short-term memory; the synapse weight storage can be equated to long-term memory.

The charge storage mechanisms used in DRAMs and EEPROMs can be utilized to retain the synapse weighting information. Electronic charge in DRAM cell is stored temporarily at the dynamic capacitor, while the charge in the EEPROM cell is stored permanently in a floating gate transistor.

4.2.1 DRAM-Style Weight Storage

The information of a synapse weighting is stored at the capacitance associated with the controlling gate in the voltage format. The synapse voltage is externally programmed through the address decoder as shown in Fig. 4.4. Since the capacitor charged by V_{ij}^s is connected to transistor M_6, leakage will occur through the diffusion-to-substrate junction. The simulated results of synapse accuracy as a function of the refresh cycle are shown in Fig. 4.6 by using a simple RC model. Since the network retrieving and synapse weighting refresh can be conducted concurrently, no dead time for refreshing the synapse voltage is required. Thus, the learning process can be realized directly. In addition, effects due to device mismatches can be compensated during the learning process.

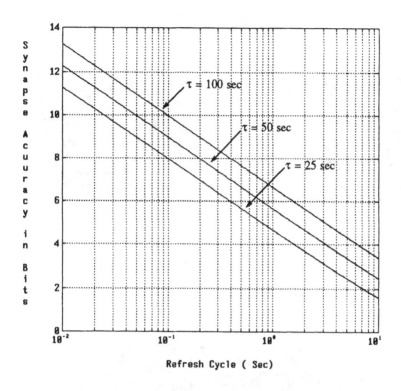

Fig. 4.6 Simulated charge retention characteristics in DRAM-style synapse cell.

The die photo of the synapse cell and charge retention characteristic of this synapse cell is shown in Fig. 4.7. The programmable synapse cell occupies an area of 46 μm x 64 μm in the Scalable 2-μm CMOS process from the MOSIS Service. In the charge retention experiment, 72 synapse cells are tied together to minimize the effect of external stray capacitance. The RC time constant is about 50 sec. Here, the x-axis is 10 sec/div and the y-axis is 0.2 V/div. A refreshing cycle of 0.2 sec is adequate for the 8-bit synapse accuracy. The measured synapse charac-teristics are shown in Fig. 4.7(c). Since a 10-to-1 resistor divider configuration is used in the test circuit, the output voltage is shown in 10x reduced format. In this experiment, the input neuron voltage of all

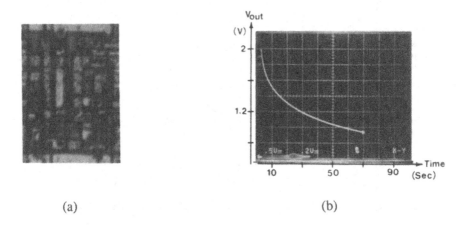

(a)   (b)

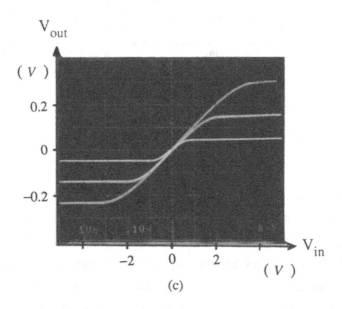

(c)

Fig. 4.7 DRAM-style synapse cell.
(a) Die photo.
(b) Measured result of charge retention test.
(c) Measured result of the synapse characteristics.

synapses are set to logic-1 and the voltages (V_{in} and ground) are applied to the gate terminals of the M_2 transistors in synapse cells. The bias currents are 1.25 µA, 5 µA, and 20 µA, respectively. The input neuron and output neuron shown in Fig. 4.8 occupy an area of 90 µm x 46 µm and 140 µm x 64 µm.

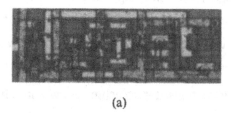

(a)

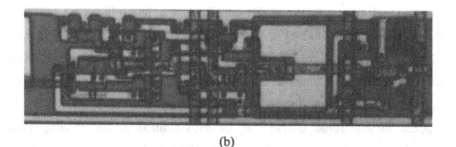

(b)

Fig. 4.8 Die photo.
(a) Input neuron.
(b) Output neuron.

4.2.2 EEPROM-Style Weight Storage

The extra refreshing circuitry using the DRAM-style storage can be saved by using the floating-gate storage approach. The usefulness of the DRAM-style storage is primarily limited by continuous charge leakage. A relatively short refresh cycle is required to compensate the charge leakage, but the high speed clocking can induce switching noise. In the EEPROM-style storage, the synapse weight information is retained in a very long time so that the learning process of the network does not have to be repeated when the electric power of the system is lost.

The floating-gate transistor structure shown in Fig. 4.9 is used to store the synapse information. A conventional double polysilicon CMOS process is used to achieve high product yield for logic ICs and good charge retention characteristics in the floating-gate layer. The polysilicon-1 (POLY1) layer is floated and the polysilicon-2 (POLY2) electrode serves as the controlling gate. The oxide thickness between the two polysilicon layers is 55 nm. Since the bump-like area enhances the local electrical field by a factor of 4 to 5, the corresponding Fowler-Nordheim tunneling voltage [20,21] in the floating-gate transistor is similar to that of a conventional EEPROM device. In addition, the tunneling voltage decreases as the number of bumps increases. The desired amount of electronic charge can be programmed to the floating layer with the external voltage applied to the POLY2 electrode.

When the programming voltage V_{prog} is applied to the POLY2 electrode, the initial value of V_{ij}^s is determined by the ratio of POLY2-to-POLY1 capacitance and POLY1-to-substrate capacitance. If the voltage drop across the POLY2-to-POLY1 capacitance is big enough to initiate the tunneling mechanism, then V_{ij}^s is exponentially changed. The equivalent model for the double-ploy transistor is shown in Fig. 4.9(c). The stored charge can be controlled with the magnitude and pulse width of V_{prog}. For a wide range and a fine resolution of synapse weightings, a small and narrow pulse width of V_{prog} is preferred. Notice that the network retrieving should be interrupted during the synapse programming period, because the magnitude of V_{prog} is higher than normal voltage for

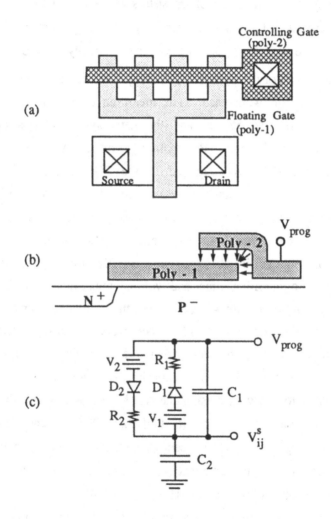

Fig. 4.9 Floating-gate transistor.
(a) Top view.
(b) Cross section.
(c) Equivalent large-signal model.

network operation. To reduce the magnitude of the programming voltage, the POLY2-to-POLY1 capacitance can be minimized with respect to the POLY1-to-substrate capacitance. The allowable magnitude of V_{prog} is limited by the pn-junction breakdown voltage between the diffusion region and substrate of transistors $M_6 - M_7$ in Fig. 4.4. These transistors are part of the decoder circuitry. In the present MOSIS 2-μm CMOS technologies, the breakdown voltage is usually around 26 V [22].

In the laboratory experiments, floating-gate transistors with W = 16 μm and L = 2 μm were used. The measured oxide thickness between polysilicon layers was 55 nm and the gate-oxide thickness under the POLY1 layer was 28 nm. Programming voltages with a fixed pulse width and magnitude were applied to the POLY2 electrode and the threshold voltage of the transistor was measured. Plots of threshold voltage changes in the fixed pulse width case and in the variable pulse width case are shown in Fig. 4.10(a) and (b), respectively. The transistor threshold voltage is changed exponentially with the programming time. Both programming time and voltage magnitude can be used as control parameters. Figure 4.10(c) shows the results of using the fixed-width pulses and increases by 0.5 V per every 10 pulses. The magnitude difference of V_{prog} for decreasing and increasing the programming voltage is caused by the MOS capacitor structure of the POLY1-to-substrate capacitor. By increasing the number of bumps and the ratio of POLY1-to-substrate capacitance to POLY2-to-POLY1 capacitance, the magnitude of V_{prog} can be reduced by a factor of more than 2 [23]. The measured output characteristics of a simple floating-gate transistor with the threshold voltage being 0.96 V and 3.45 V are shown in Fig. 4.11(a) and (b), respectively. The transistor drain current was measured with the gate voltage changed from 1 V to 7 V in steps of 1 V. High-quality transistor output characteristics can be observed in both cases.

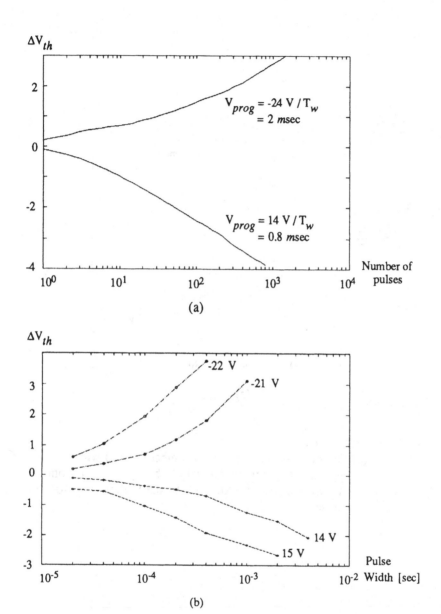

Fig. 4.10   Measured results of several programming schemes.
(a) Threshold voltage change versus the number of
fixed-width/magnitude pulses.
(b) Threshold voltage change versus a single pulse
with variable width.

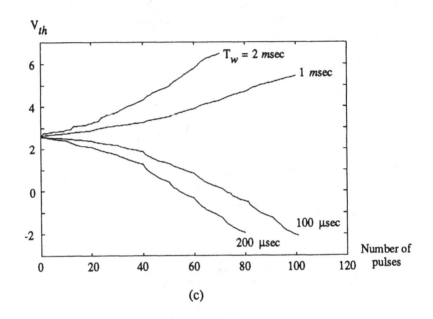

(c)

Fig. 4. 10 (Continued)
　　　　(c) Threshold voltage versus the number of fixed-width
　　　　　　pulses with programming voltage being increased from
　　　　　　12 V or decreased from -20 V by 0.5 V after ecery 10
　　　　　　pulses.

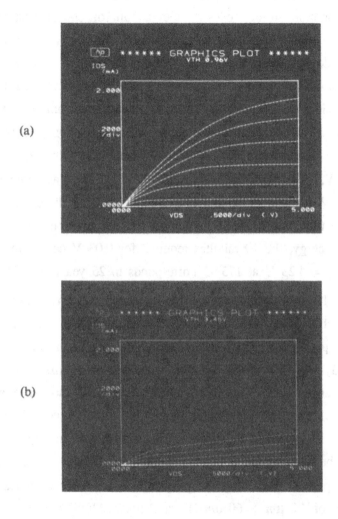

Fig. 4.11   Output characteristics of a double-polysilicon floating-gate
transistor with W/L = 18 μm / 2 μm. The gate voltage changes
from 1 V to 7 V in step of 1V.
(a) The threshold voltage is programmed to be 0.96 V.
(b) The threshold voltage is programmed to be 3.45 V.

Charge retention characteristic is of primary importance in long-term information storage because the amount of charge leakage determines the ultimate accuracy of the circuit. In the experiments, programmed devices were baked at $175°C$ and $225°C$ to accelerate the charge relaxation process. Figure 4.12 shows the sustained threshold voltage against storage time for devices with different amounts of charge accumulation in the floating-gate layer. The activation energy plot of charge loss is shown in Fig. 4.12(b) by using data from $V_{th} = 4.23$ V and 2.91 V. The higher activation energy of 1.51 eV as compared with the 1.10 eV for critical-oxide EEPROM devices is a direct benefit from the thicker tunneling oxide in the simple floating-gate transistors. At this activation energy, the 10 minutes required for 0.03 V charge loss for the case of $V_{th} = 4.23$ V at $175°C$ corresponds to 25 years at $55°C$.

The 10 mV resolution can be achieved by single programming pulse of 100 μsec at $V_{prog} = 12$ V and -20 V. In the case of $I^{max} = 20$ μA and ± 5 V-power supply, the maximum V_{limit} is 3.2 V and the dynamic range for synapse weighting is 50 dB. The synapse dynamic range will be increased as the magnitude and pulse width of V_{prog} decreases. The die photos of a 4-bit neural-based A/D converter using the EEPROM-style storage cells are shown in Fig. 4.13. To increase the number of bumps, a finger-type POLY1 pattern is used. Each synapse cell including the programming circuitry occupies a compact area of 70 μm $\times$ 60 μm in a 2-μm CMOS process, which is almost 10 times that of a SRAM cell. Thus, a VLSI neural chip with 100k programmable synapses can be manufactured using the one megabit CMOS process for memory products.

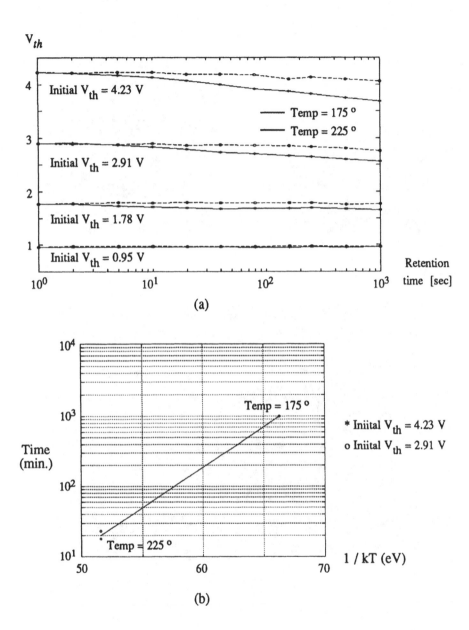

Fig. 4.12 Charge retention characteristics of a simplie floating-gate
transistor.
(a) Sustained threshold voltage against storage time for
175 ° and 225 °.
(b) Activation energy plot of the charge loss.

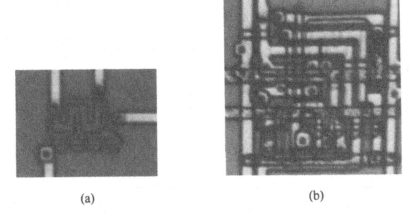

(a)   (b)

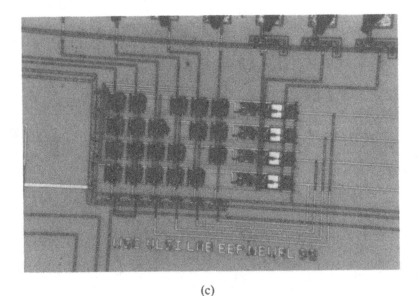

(c)

Fig. 4.13 Die photo.
     (a) Programming cell.
     (b) EEPROM-style synapse cell.
     (c) 4-bit Hopfield A/D converter with EEPROM-
         style synapse cell.

References

[1]   C. A. Mead, *Analog VLSI and Neural Systems*, New York: Addison-Wesley, 1989.

[2]   R. Hecht-Nielsen, "Neural-computing: picking the human brain," *IEEE Spectrum*, vol. 25, no. 3, pp. 36-41, Mar. 1988.

[3]   R. Hecht-Nielsen, *Neurocomputing*, New York: Addison-Wesley, 1990.

[4]   B. W. Lee, B. J. Sheu, *Design and Analysis of VLSI Neural Networks*, in Neural Networks: Introduction to Theory and Applications, Editor: B. Kosko, Englewood Cliffs, NJ: Prentice-Hall, 1990.

[5]   D. E. Van den Bout, P. Franzon, J. Paulos, T. Miller III, W. Snyder, T. Nagle, W. Liu, "Scalable VLSI implementations for neural networks," *Jour. of VLSI Signal Processing*, vol. 1, no. 4, pp. 367-385, Boston, MA: Kluwer Academic, Apr. 1990.

[6]   P. W. Hollis, J. J. Paulos, "Artificial neural networks using MOS analog multipliers," *IEEE Jour. of Solid-State Circuits*, vol. 25, no. 3, pp. 849-855, June 1990.

[7]   D. Hammerstram, "A VLSI architecture for high-performance, low cost, on-chip learning," *Proc. of IEEE/INNS Joint Conf. on Neural Networks*, vol. II, pp. 537-544, San Diego, CA, June 1990.

[8]   M. Holler, S. Tam, H. Castro, and R. Benson, "An electrically trainable artificial neural network (ETANN) with 10240 'float gate' synapses," *Inter. Joint Conf. on Neural Networks*, vol. 2, pp. 191-196, June 1989.

[9]   T. Morishita, Y. Tamura, and T. Otsuki, "A BiCMOS analog neural network with dynamically updated weights," *IEEE Inter. Solid-State Circuits Conf.*, pp. 142-143, Feb. 1990.

[10]   B. W. Lee, B. J. Sheu, "A compact and general-purpose neural chip with electrically programmable synapses," *Proc. of IEEE Custom Integrated Circuits Conf.*, pp. 26.6.1-26.6.4, Boston, MA, May 1990.

[11]   D. E. Van den Bout and T. K. Miller III, "A digital architecture employing stochasticism for the simulation of Hopfield neural nets," *IEEE Trans. on Circuits and Systems,* vol. 36, no. 5, pp. 732-738, May 1989.

[12]   T. Quarles, SPICE3 Version 3C1 Users Guide, *Electron. Res. Lab. Memo UCB/ERL M89/46*, University of California, Berkeley, Apr. 1989.

[13]   A. Vladimirescu, S. Liu, "The simulation of MOS integrated circuits using SPICE2," *Electron. Res. Lab. Memo ERL-M80/7,* University of California, Berkeley, Oct. 1980.

[14]   B. J. Sheu, D. L. Scharfetter, P. K. Ko, M.-C. Jeng, "BSIM: Berkeley short-channel IGFET model for MOS transistors," *IEEE Jour. of Solid-Stae Circuits,* vol. SC-22, no. 4, pp. 558-566, Aug. 1987.

[15]   HSPICE Users' Manual H9001, Meta-Software Inc., Campbell, CA, May 1989.

[16]   R. E. Howard, D. B. Schwartz, J. S. Denker, R. W. Epworth, H. P. Graf, W. E. Hubbard, L. D. Jackel, B. L. Straughn, and D. M. Tennant, "An associative memory based on an electronic neural network architecture," *IEEE Trans. on Electron Devices,* vol. ED-

34, no. 7, pp. 1553-1556, July 1987.

[17] B. W. Lee and B. J. Sheu, "Design of a neural-based A/D converter using modified Hopfield network," *IEEE Jour. of Solid-State Circuits*, vol. SC-24, no. 4, pp. 1129-1135, Aug. 1989.

[18] B. W. Lee, B. J. Sheu, "A high slew-rate CMOS amplifier for analog signal processing," *IEEE Jour. of Solid-State Circuits*, vol. 25, no. 3, pp. 885-889, June 1990.

[19] B. W. Lee, B. J. Sheu, J. Choi, "Programmable VLSI neural chips with hardware annealing for optimal solutions," *IEEE Jour. of Solid-State Circuits*, to appear.

[20] R. K. Ellis, "Fowler-Nordheim emission from non-planar surfaces," *IEEE Electron Device Letters*, vol. 3, no. 11, pp. 330-332, Nov. 1982.

[21] S. M. Sze, *Physics of Semiconductor Devices*, pp. 500-504, 2nd Edition, New York: John Wiley & Sons, 1981.

[22] C. Tomovich, "MOSIS: A gateway to silicon," *IEEE Circuits and Devices Magazine*, vol. 4, no. 2, pp. 22-23, Mar. 1988.

[23] H. A. R. Wegener, "Endurance model for textured-poly floating gate memories," *IEEE Inter. Electron Devices Meeting*, pp. 480-483, Dec. 1984.

Chapter 5

System Integration for VLSI Neurocomputing

VLSI neural systems play an important role in real-time signal and image processing. Several neural hardware designs which typically function as direct accelerators for software computation have been reported [1]. These hardware accelerators, including main processors, mathematical coprocessors, and associated memories, solve difference equations of a neural network instead of the original differential equations. The approximation of using difference equations requires a tremendously large computational time and also could lead to a fatal failure caused by numerical errors [2]. The rich properties of neural networks inherent in the massively parallel processing scheme using programmable synapses and neurons are to be fully explored. To make the neurocomputing hardware more powerful, compact and electrically programmable synapses are crucial. Using the programmable neural chip with gain-adjustable neuron and adaptive synapses, reconfigurable neural systems with learning capability can be produced. In addition, hardware annealing can easily be applied to obtain the optimal solutions.

5.1 System Module Using Programmable Neural Chips

One good example of using the DRAM-style VLSI programmable neural network is shown in Fig. 5.1. The synapse weight data are stored in the buffer memory and are used to dynamically refresh the on-

chip synapse voltage through a D/A converter. The host computer
sends the synapse weight data into the buffer memory, which consists of
two-port SRAMs, through the interface circuitry. The memory address
for the refresh operation is generated by the clock generator and counter.
The network learning process is simply done by writing synapse information into the buffer memory without stopping the retrieving process.

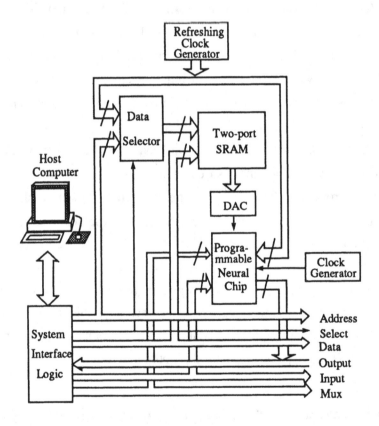

Fig. 5.1 Neural computing board diagram using DRAM-style
programmable synapses.

Thus, the network retrieving and learning processes can be performed concurrently so that the back-propagation learning scheme can easily be implemented. The input and output signals for network retrieving are communicated with host computer according to the multiplexing signals. Notice that the analog multiplexers inside the VLSI chip are used for VLSI neural chips with high numbers of I/O's. The synapse accuracy can easily be enhanced by decreasing the refresh time and increasing the number of bits in external memory and the D/A converter. The analog multiplexers were also used in the design of VLSI components for a general purpose analog neural computer by P. Mueller, et al. [3,4].

The use of EEPROM-style programmable neural chips can save the external synapse weight memories and the refresh circuitry. Figure 5.2 shows the system-level block diagram for EEPROM-style design. During the learning process, the host computer sends out address signals and synapse weight data. At the high-voltage pulse generator, the synapse

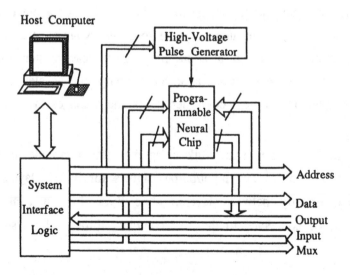

Fig. 5.2 Neural computing board diagram using EEPROM-
style programmable synapses.

weight data is converted to the corresponding pulse signals with appropriate pulse width and magnitude. Since the synapse weight programming voltage is usually very high, the learning process should be separated from the network retrieving process. In a supervised learning case, the learning process takes a relatively long time because network outputs are monitored at each learning cycle. In addition, a certain waiting period after each programming cycle is also required due to charge relaxation in the oxide layer. However, the excellent charge retention characteristics of the EEPROM-style neural chips can certainly save the large overhead for synapse weight refresh [5,6].

The system module including the programmable neural chip is reconfigurable with external connections. Even though each module has a fixed number of neurons and synapses, any neural network architectures including the Hopfield network, bidirectional associative memory, and multilayer network can be built with multiple copies of such modules. Detailed review of neural networks architectures for efficient hardware implementations can be found in [7-13]. For a practical application with a very large number of neurons and synapses, the computing job can be partitioned into an appropriate size for each system module to handle. The module functions as a neural computing accelerator and the host computer or a system controller stores and manages the intermediate data.

A multilayer neural network using the system module approach is shown in Fig. 5.3. The input/output signals for hidden layers can be connected directly between layers, while those for input and output layers can be connected to the host computer through the system interface logic.

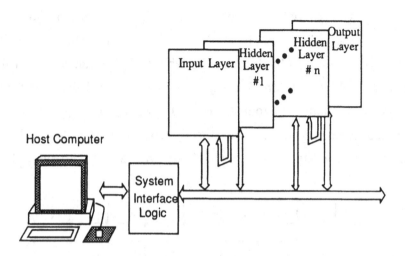

Fig. 5.3 Real-time neural computing system with a multi-layer network

5.2 Application Examples

Two VLSI neural chips which use the programmable synapses and gain-adjustable neurons are demonstrated. One chip with only analog circuitry operates asynchronously, while the other chip with mixed analog-digital circuitry operates synchronously. The computational time of the special neural hardware is also compared with that of the corresponding software computation on standard workstations.

5.2.1 Hopfield Neural-Based A/D Converter

The adaptive synapses and gain-adjustable neurons can make neural-based A/D converters more flexible in the output characteristics.

Since many synapses have the same weighting, the voltage can be shared with many synapse cells, as shown in Fig. 5.4. Figure 5.5 shows the die photo of an 8-bit neural-based A/D converter. The chip size is 800 μm x 550 μm in the MOSIS 2-μm Scalable CMOS technology. In this design, the values of feedback synapses are negative and the neuron outputs are in a noninverted format. For synapses T_{iS} and T_{iR}, the analog input and reference voltage are directly applied to the terminals for synapse weight voltage storage, while the synapse bias currents (I_j) are

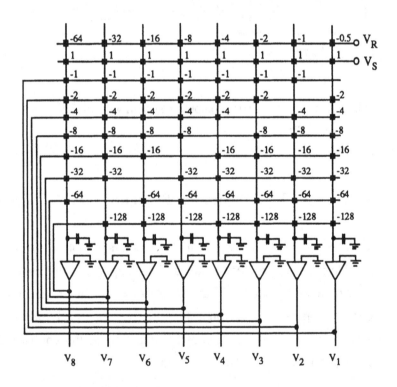

Fig. 5.4   Circuit schematic of an 8-bit neural-based A/D converter with the programmable synapses and gain-adjustable neurons.

always turned on. Thus, the synapse weight voltage of the reference synapses is also shared with that of the feedback synapses. Notice that the negative synapse weight can be obtained by applying a negative voltage to the weight storage capacitance. When the synapse weight voltages are derived from the R-2R ladder circuit and external voltage V_{ext}, the conversion step size is $V_{ext}/2^n$. The rich local minima, which appear as the overlapped analog input range in the transfer function, can help to compensate the initial device mismatches. By applying the hardware annealing technique, the local minima can be eliminated so that the transfer curve becomes pretty linear.

Fig. 5.5 Die photo of the 8-bit neural-based A/D converter
in Fig. 5.4.

The conversion speed of the original Hopfield A/D converter and the self-correction A/D converter are shown in Fig. 5.6. In this experiment, the 4-bit neural-based A/D converter is used. The response time can be grouped easily into four classes, which are dependent upon the number of convergence iterations. Notice that there are some missing codes in the Hopfield A/D converter because of the local minima. No significant change of the response time in the self-correction A/D converter is observed. Here, the unity-gain bandwidth of the neuron amplifier is about 1 MHz and the response time with a pulsed input signal of 2 V_{p-p} is approximately 1.5 µsec. On the other hand, delay of the correction logic is approximately 10 nsec.

The transient characteristics of the individual amplifier are shown in Fig. 5.7. In the transition from (0000) to (0111), the four iterations to settle down the final state are clearly displayed. In another case using a 5-bit Hopfield A/D converter, 5 iterations and 4 iterations are required between transitions (00001)-to-(01101) and (01101)-to-(00001), respectively. This experimental result shows that the maximum response time of a Hopfield neural-based A/D converter is determined by the number of converter bits. Notice that the time delay of an off-the-shelf D/A converter, which is used to reconstruct the A/D converter output, is 1 µsec.

The speed comparison between the IC implementation and a software computation is listed in Table 5.1. Here, the total computational time is defined as the amount of time to obtain full-scaled increasing and decreasing analog inputs with 300 evaluation points. In software computation, the Hopfield differential equations are solved by the forward-Euler integration method with fixed time steps and with adaptive time steps. For the adaptive time-step case, a local numerical error is estimated with a Richardson extrapolation method [14]. The program was written in C-language and executed in SUN 4/60 work-

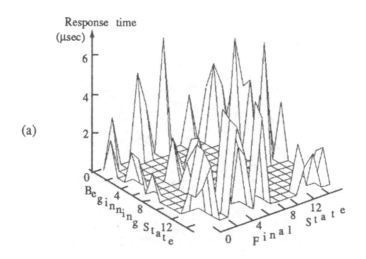

(a)

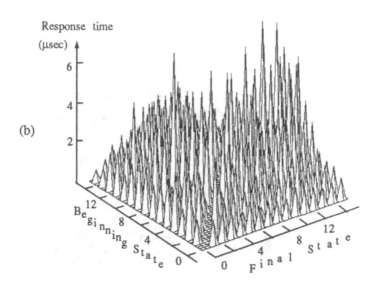

(b)

Fig. 5.6 Response time characteristics of 4-bit neural-based
    converters.
    (a) Original Hopfield A/D converter.
    (b) Self-correction A/D converter.

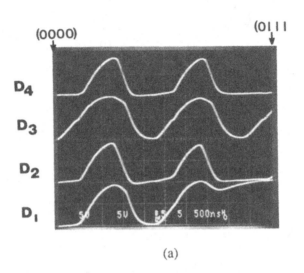

(a)

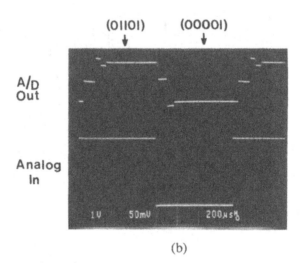

(b)

Fig. 5.7 Transient response.
        (a) Each amplifier output of a Hopfield 4-bit A/D converter.
        (b) 5-bit Hopfield A/D converter.

Table 5.1 Computational speed of 4-bit A/D conversion

Method / Item	Hardware Computation using IC	Software Computation			
		Adaptive time step	Fixed-time step		
			1.x 10^{-2}	1.x 10^{-3}	1 x 10^{-4}
Computational Time (sec)	1.8 x 10^{-3}	314	182	1697	16147
CPU-time factor	1	1.7x 10^5	1.0x 10^5	9.4x 10^5	9.0 x 10^6
Quality of Solution	Good	Good	Bad	Poor	Good

stations. For the speed calculation of the IC implementation, the worst-case conversion time (6 μsec) for one conversion was used. The initial condition for each analog input voltage was reset in software computation. The simulated transfer curve for the time step of 1.0×10^{-4} sec is the same as that for the adaptive time-step case, as shown in Fig. 5.8. Notice that the adaptive time-step scheme is equivalent to solving the original differential equations, while the fixed time scheme is equivalent to solving the approximated difference equations. As compared to the adaptive time-step scheme, hardware computing enhances the speed by a factor of 17,400.

In the VLSI implementation, amplifier offset voltage and synapse weight tolerance caused by device mismatches are the major limitations. When the amplifier input offset voltage V_{ios} exists, it affects the decision result for the output neuron. The neuron input voltage is determined by

$$u_i = \frac{\sum\limits_{j=1, j \neq i}^{n} T_{ij} V_j + T_{iS} V_S + T_{iR} V_{iR}}{T_i} , \qquad (5.1)$$

where T_i is total input conductance to the i-th neuron. The shift of the analog input voltage is given by

$$\Delta V_S = \frac{T_i}{T_{iS}} \times V_{ios} \ . \tag{5.2}$$

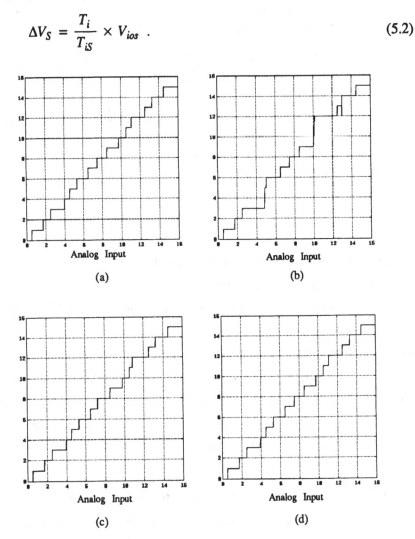

Fig. 5.8 Software simulation results.
    (a) Adaptive time step.
    (b) Fixed time step of 1.0×10^{-2} sec.
    (c) Fixed time step of 1.0×10^{-3} sec.
    (d) Fixed time step of 1.0×10^{-4} sec.

Here, V_{ios} is the amplifier input offset voltage. The shift of the analog voltage range is magnified by the ratio of T_i/T_{iS}. When the number of neurons, n, is very large, the asymptotic value for the magnification factor is

$$\frac{T_i}{T_{iS}} \approx \begin{cases} n + 1 & \text{with programmable synapse} \\ 2^n & \text{with direct–R synapse .} \end{cases} \tag{5.3}$$

Notice that actual conductance value in the DRAM-style synapse is constant, while the synapse weight is determined by the external voltage. Thus, the input voltage shift of the programmable synapse is much smaller than that of the direct-resistance synapse.

The tolerance of the synapse weight also affects the analog input voltage range. When V_{ios} is very small and the amplifier gain is very large, the analog input voltage is determined by (2.39) and (2.40). Thus, the change of the analog input voltage range caused by the synapse weighting tolerance is given as

$$\delta V_S = - \frac{\sum\limits_{j=1, j \neq i}^{n} V_j(\delta T_{ij}) - V_R(\delta T_{iR})}{T_{iS}} + \frac{\delta T_{iS}}{T_{iS}} \times \frac{\sum\limits_{j=1, j \neq i}^{n} T_{ij} V_j + T_{iR} V_{iR}}{T_{iS}} \tag{5.4}$$

Here, δT_{ij}, δT_{iS} and δT_{iR} are the tolerances of T_{ij}, T_{iS}, and T_{iR}, respectively. Since the absolute value of the tolerance usually increases with the synapse conductance value in the VLSI technology, the change in analog input range of a direct-R implementation is much larger than that of the DRAM-style implementation.

The measurement results of a Hopfield neural-based A/D converter whose synapses were realized with resistors using p-well regions are shown in Fig. 5.9. In this design, the T_{iS} is scaled up by 10 times and the analog input voltage range is reduced by 10 times. When the bias current for the amplifier increases from 2.5 μA to 12.5 μA, the analog input range is shifted by approximately 30 mV due to the increase of the amplifier offset voltage. The output neuron is a simple 2-stage CMOS amplifier and the A/D converter output is reconstructed with a conventional D/A converter. Notice that one neuron output at the boundary value of the analog input voltage range is usually not a discrete value due to the finite amplifier gain. Thus, a small overlapped input range for digital codes (0000) and (0001) can exist due to the A/D output reconstructing process.

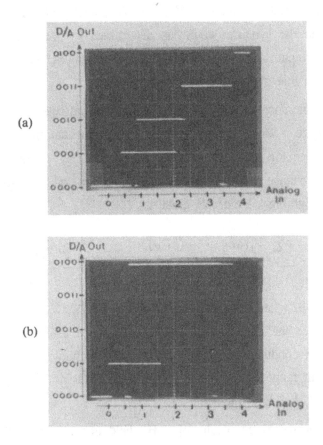

Fig. 5.9 Finite gain and input offset voltage effects.
(a) Bias current for each amplifier is 2.5 μA.
(b) Bias current for each amplifier is 12.5 μA.

The die photos of two general purpose chips using DRAM-style synapse cells are shown in Fig. 5.10. The chip sizes are 4400 μm x 6600 μm and 2500 μm x 2350 μm in a MOSIS 2-μm CMOS technology, respectively. Each chip includes a single layer network with fully connected synapses. The input/output pads of the chip in Fig. 5.10(a) are multiplexed in time in order to solve the limited number of package pins. Thus, the chip including 64 input neurons, 64 output neurons, and 4096 programmable synapses could operate in the synchronous mode. Conversely, the other chip, shown in Fig. 5.10(b), including only 25 input neurons, 12 output neurons, and 300 synapses, can operate in the asynchronous mode.

5.2.2 Modified Hopfield Network for Image Restoration

Image restoration is an important task in digital image processing for machine intelligence [15-19]. The image quality is recovered from a degraded image caused by optical system aberration, motion, diffraction, and noise. The image restoration process, which finds the improved image $\mathbf{X}$ by minimizing the error function E with respect to the degraded image $\mathbf{Y}$, can be described as follows [15],

$$E = \frac{1}{2}\|\mathbf{Y} - \mathbf{HX}\| + \frac{1}{2}\lambda\|\mathbf{DX}\| , \qquad (5.5)$$

where $\| . \|$ is the L_2 norm, $\mathbf{H}$ is the blur matrix, λ is a constant, and $\mathbf{D}$ is the sharpness matrix. In the case that the image has L x L pixels with M-bit gray levels, the error function in (5.5) can be described as a Hopfield energy function. The synapse weightings $\{T_{i,k;j,l}\}$ and the input currents $\{I_{i,k}\}$ of a Hopfield network are given as [20]

$$T_{i,k;j,l} = - \sum_{p=1}^{L^2} h_{p,i}h_{p,j} - \lambda \sum_{p=1}^{L^2} d_{p,i}d_{p,j} \qquad (5.6)$$

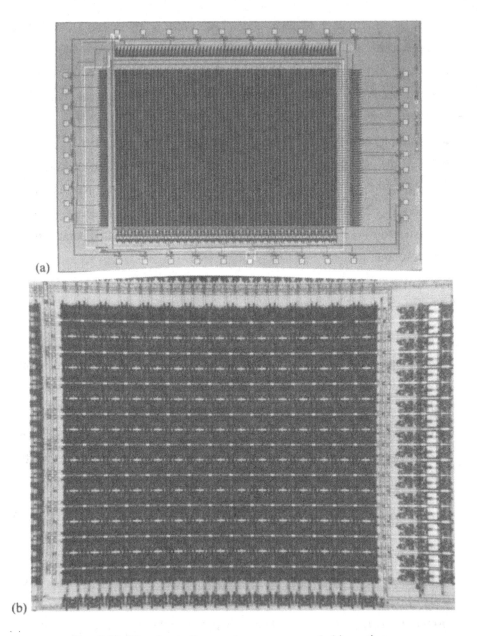

(a)

(b)

Fig. 5.10 Die photo of general-purpose neural chips using
          DRAM-style storage.
          (a) With multiplexing I/O pins.
          (b) Without multiplexing I/O pins.

and

$$I_{i,k} = \sum_{p=1}^{L^2} y_p h_{p,i} \; . \tag{5.7}$$

Here, i, j, k, and l are integers in $[1, L]$.

Since the self-feedback terms of synapse matrix $\{T_{i,k;j,l}\}$ are not zero, the Hopfield network will not always move toward a lower energy level. In addition, $L^2 M$ neurons and $L^4 M^2/2$ synapses are required for the Hopfield network. Detailed algorithms for image restoration using Hopfield networks can be found in [20]. In a modified Hopfield network, the number of neurons and synapses are reduced to L^2 and L^4, respectively. Each output neuron generates an increment/decrement signal for the M gray-level register in a synchronous manner. During iterations, the gray level for each pixel is increased or decreased by the pixel neuron output. The initial states for each gray level register are from the blurred image.

The chip-level architecture for the real-time image restoration is shown in Fig. 5.11. For VLSI implementation, the whole image is partitioned into subimages of n x n pixels and the subimage is processed in one VLSI chip. The n^2 neuron outputs are sent to the pixel registers. Each pixel register consists of an M-bit counter which stores the pixel gray level. The initial content of pixel registers is loaded from the blurred image and the final content of pixel registers is the restored image. Notice that the D/A conversion function for the pixel register is effectively realized in the binary weighed synapses. Thus, one synapse value in (5.6) is coded into M synapses $T_{i,k;j,l}^r$.

$$T_{i,k;j,l}^r = T_{i,k;j,l} \times 2^r \; , \tag{5.8}$$

where r is $[1, M]$. The synapse weightings of the fully connected n^4 x M synapses are determined from parameters of the blur function and smoothing constraint, while that of the n^2 synapses are determined from

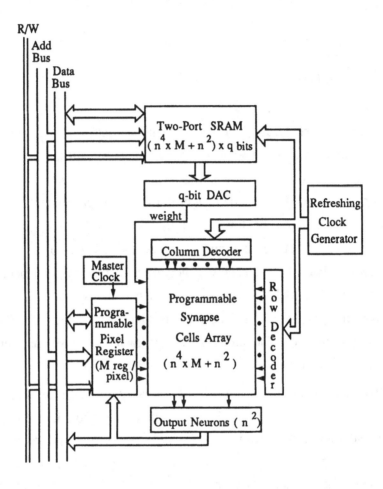

Fig. 5.11 Architecture of image processing neural chip.

the input currents. Although the synapse matrix is usually very sparse, the fully connected synapse matrix is implemented to examine various blur and smooth functions. Two-port SRAM with $(n^4 \times M + n^2) \times q$ bits are used for the learning process and the retrieving process. The synapse weighting is described in q-bit accuracy. In typical applications, $q = 8$ is used. The refresh address generator provides the address signals to row/column decoder and the SRAM. The synapse weight data in the SRAM are sent to the synapse cells. Figure 5.12 shows the 1-bit digital circuit cell for the pixel register, which includes the read/write circuitry and counter. The die photo of a modified Hopfield network for processing 5x5-pixel subimages is shown in Fig. 5.13. In this design, a 4-bit resettable counter per one pixel is used to restore an image with 16 gray levels. Thus, the chip includes 25 gain-adjustable neurons and 2525 programmable synapses in 6800 μm x 4800 μm chip size using the MOSIS 2-μm Scalable CMOS technology.

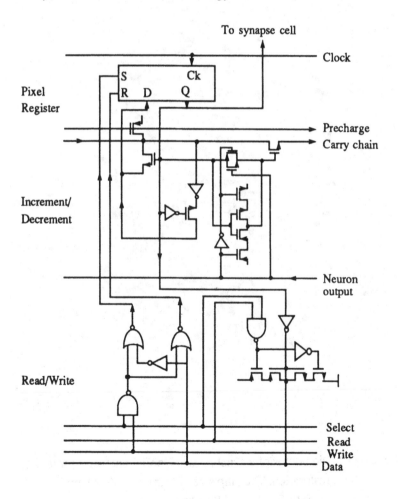

Fig. 5.12 Logic schematic of one pixel programmable register.

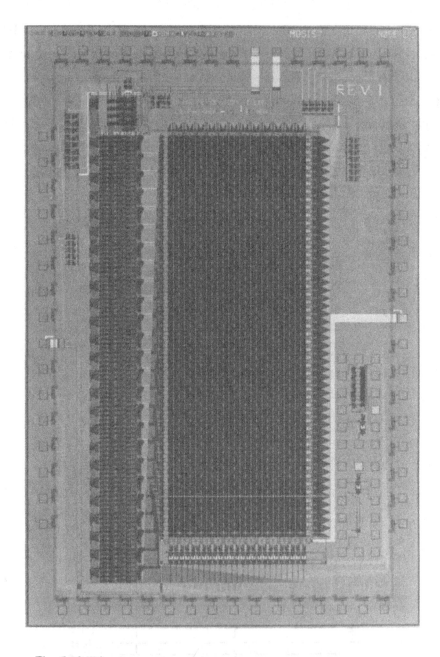

Fig. 5.13 Die photos of a synchronous Hopfield network for image
restoration. The chip size is 4600 μm x 6800 μm in
MOSIS 2-μm CMOS technology.

System level simulation using a girl image and a house image of 256 x 256 pixels has been conducted to evaluate the performance of the image restoration chip [21]. The images were blurred by a uniform and shift-invariant blur function of a 3x3 window size. The blurred function **H** used in this simulation is

$$\mathbf{H} = \begin{bmatrix} 1/9 & 1/9 & 1/9 \\ 1/9 & 1/9 & 1/9 \\ 1/9 & 1/9 & 1/9 \end{bmatrix}. \tag{5.9}$$

Then, the synapse weighting in (5.8) is given as

$$T_{i,k;j,l} = \begin{cases} -9/81 & \text{if } i=j \cap k=l \\ -6/81 & \text{if } (|i-j|=1 \cap k=l) \cup (i=j \cap |k-l|=1) \\ -4/81 & \text{if } |i-j|=1 \cap |k-l|=1 \\ -3/81 & \text{if } (|i-j|=2 \cap k=l) \cup (i=j \cap |k-l|=2) \quad (5.10) \\ -2/81 & \text{if } (|i-j|=2 \cap |k-l|=1) \cup (|i-j|=1 \cap |k-l|=2) \\ -1/81 & \text{if } |i-j|=2 \cap |k-l|=2 \\ 0 & \text{otherwise}. \end{cases}$$

Notice that the maximum number of neurons whose outputs are not discrete values is 6 in this blurred function. The image is partitioned into 5 x 5 subimages and processed with the VLSI chip architecture of the modified Hopfield network. The image restoration process is performed for 20 iterations. The blurred and restored images are shown in Fig. 5.14. The performance comparison of adaptive neural chips with a Sun-3/60 workstation in image restoration is listed in Table 5.2. The speedup factor using a single chip of 25 neurons is 96. Further improvement can be achieved with the use of an array of neural chips as shown in Fig. 5.15.

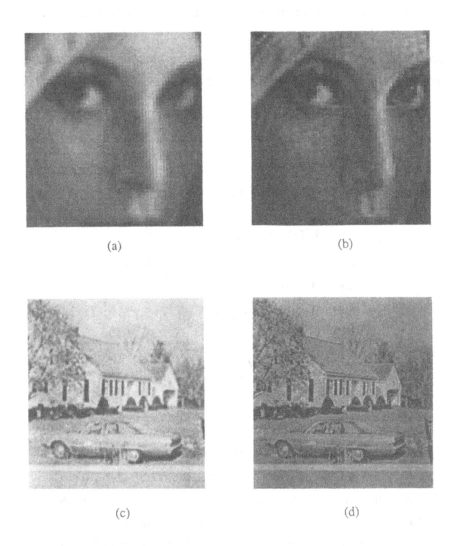

(a)   (b)

(c)   (d)

Fig. 5.14 System simulation results on the images.
(a) Blurred picture of girl image.
(b) Restored picture of girl image.
(c) Blurred picture of house image.
(d) Restored picture of house image.
The picture is partitioned into 5 x 5 subimages and
processed using the synchronous Hopfield chip architecture.

Table 5.2 Speed performance of the neural-based image restoration chip.

(An image with 256x256 pixels and 16 gray levels per pixel is used.)

System / Performance	25-neuron chip
Synapse Programming time	27 ms
Pixel register read/write time	880 ms
Network execution time w/ 20 iter./subimage	2601 ms
Total processing time for whole image	3.51 sec
Speed-up factor to Sun-3/60	341

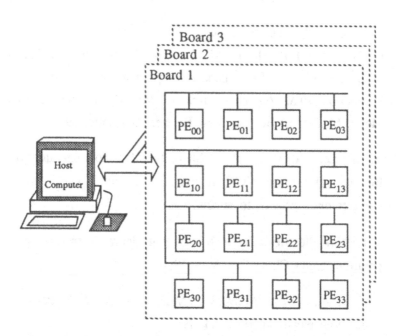

Fig. 5.15 Schematic diagram of a multi-chip image restoration systems.

References

[1] R. Hecht-Nielsen, "Neural-computing: picking the human brain," *IEEE Spectrum*, vol. 25, no. 3, pp. 36-41, Mar. 1988.

[2] B. W. Lee and B. J. Sheu, "Combinatorial optimization using competitive-Hopfield neural networks," *Proc. of Inter. Joint Conf. on Neural Networks*, vol 2, pp. 627-630, Washington D.C., Jan. 1990.

[3] P. Mueller, J. Van der Spiegel, D. Blackman, T. Chiu, T. Clare, J. Dao, C. Donham, J.-P. Hsieh, M. Loinaz, "A general purpose analog neural computer," *Proc. of IEEE/INNS Inter. Conf. on Neural Networks*, vol. II, pp. 177-182, Washington D.C., June 1989.

[4] P. Mueller, J. Van der Spiegel, D. Blackman, T. Chiu, T. Clare, C. Donham, J.-P. Hsieh, M. Loinaz, "Design and fabrication of VLSI componenets for a general purpose analog neural computer," in *Analog VLSI Implementation of Neural Systems*, Editors: C. Mead, M. Ismail, Boston, MA: Kluwer Academic, 1989.

[5] B. W. Lee, H. Yang, B. J. Sheu, "Analog floating-gate synapses for general-purpose VLSI neural computation," *IEEE Trans. on Circuits and Systems*, to appear.

[6] 80170NW Electricaly Trainable Analog Neural Network, Intel Corporation, Santa Clara, CA, May 1990.

[7] J. Dayhoff, *Neural Network Architectures: An Introduction*, New York: Van Nostrand Reinhold, 1990.

[8] B. Soucek, M. Soucek, *Neural and Massively Parallel Computers*, New York: Wiley-Interscience: 1988.

[9]   B. Soucek, *Neural and Concurrent Real-Time Systems*, New York: Wiley-Interscience, 1989.

[10] S. F. Zornetzer, J. L. Davis, C. Lau, Editors, *An Introduction to Neural and Electronic Networks*, New York: Academic Press, 1990.

[11] Y.-H. Pao, *Adaptive Pattern Recognition and Neural Networks*, New York: Addison Wesley, 1989.

[12] R. Hecht-Nielsen, *Neurocomputing*, New York: Addison-Wesley, 1990.

[13] S. T. Toborg, K. Hwang, *Exploring Neural Network and Optical Computing Technologies*, in Parallel Processing for Supercomputers and Artificial Intelligence, Editors: K. Hwang and D. Degroot, New York: McGraw Hill, 1989.

[14] G. Dahlquist and A. Bjorck, *Numerical Methods*, pp. 330-350, Prentice-Hall Inc., 1974.

[15] H. C. Andrews and B. R. Hunt, *Digital Image Restoration*, Englewood Cliffs, NJ: Prentice-Hall Inc., 1977.

[16] W. K. Pratt, *Digital Image Processing*, New York: Wiley-Interscience, 1978.

[17] F. M. Wahl, *Digital Image Signal Processing*, Boston, MA: Artech House, 1987.

[18] A. Rosenfeld, A. C. Kak, *Digital Picture Processing*, 2nd Edition, New York: Academic Press, 1982.

[19] A. K. Jain, *Fundamentals of Digital Image Processing*, Englewood Cliffs, NJ: Prentice Hall, 1988.

[20] Y. Zhou, R. Chellappa, A. Vaid, and B. Jenkins, "Image restoration using a neural network," *IEEE Trans. Acoustics, Speech & Signal Proc.*, vol. 36, pp. 141-1151, July 1988.

[21] J.-C. Lee, B. J. Sheu, "Parallel digital image restoration using adaptive VLSI neural chips" *Proc. of IEEE Inter. Conf. on Computer Design*, Cambridge, MA, Sept. 1990.

Chapter 6

Alternative VLSI Neural Chips

The first neural hardware built by M. Minsky and D. Edmonds in 1951 was composed of simple discrete devices such as vacuum tubes, motors, and manually adjusted resistors. The machine successfully demonstrated the learning capability of a Perceptron. The Madaline/Adaline [1] was applied to the first commercial product (Memistor) which can be used for pattern recognition and adaptive control applications. Conversely, present neurocomputing machines are composed of VLSI chips and electronic storage elements. The first general-purpose neurocomputing machine, Mark III, works with a VAX mini-computer to accelerate neural processing in software computation [2]. It was reported that the simulation speed of the integrated VAX-MARK machine can be improved by approximately 29 times than that of a VAX computer alone. The ANZA and Delta-1 printed-circuit boards are accelerators for IBM PC/AT personal computers. They are composed of several VLSI chips which function as the CPU, mathematical co-processor, and memories. To make the neurocomputing hardware more powerful, design and fabrication of special VLSI neural chips are highly needed.

Several neural circuits have been developed which implement the neural networks in VLSI chips as shown in Table 6.1. Chip area and the number of package pins are major limiting factors. Such constraints can be greatly relaxed if a locally connected neural network is used in the implementation [3]. The required chip area will be gradually decreased with the advances in the CMOS fabrication technologies and with the novel circuit design techniques. On the other hand, the number of package pins which is determined by specific applications still remains the fundamental limitation. Each physical package pin can be shared by several functional I/O's through the time-domain multiplexing scheme.

Table 6.1 Reported VLSI neural chips

Developer	Complexity	Technology	Applications
W. Hubbard, et al. (Bell Lab.)	22 neurons 484 resistors	CMOS Amorphous-Si	Content Address Memory(CAM)
H. P. Graf, et al. (Bell Lab.)	256 neurons 130k resistors	2.5μm CMOS Amorphous-Si (5700μmx5700μm)	CAM, Data Compressor
M. Sivilotti, et al. (Caltech)	analog MOS circuitry 100k Trs	2.5μm CMOS	Retina
B. Lee & B. Sheu (USC)	4 neurons 23 synapses	3.0μm CMOS (2300μmx3400μm)	Neural-based A/D converter
M. Holler, et al. (Intel)	64 neurons 8192 synapses	1.0μm E^2PROM CMOS (8200μmx6400μm)	General-purpose (EEPROM)
T. Morishita, et al. (Matsushita)	64 neurons 768 synapses	2.2μm BiMOS (18000μmx13500μm)	General-purpose (DRAM)
H. P. Graf , et al. (Bell Lab.)	256 neurons 32k synapses	0.9μm CMOS (4500μmx7000μm)	Reconfigurable General-purpose (SRAM)
B. Lee & B. Sheu (USC)	64 neurons 4096 synapses	2.0μm CMOS (4600μmx6800μm)	General-purpose (DRAM)

Various circuit design styles including pure analog, pure digital, and mixed-signal circuitries and different fabrication technologies including MOS, CCD, BiCMOS, and optoelectronic processes have been used to construct artificial neural chips [4-16]. In the following sections, several chip architectures and design styles will be examined in detail.

6.1 Neural Sensory Circuit

Neural sensory circuits directly convert real-world signals including images and sounds into internal neural representations. Since the real-world signal is usually in the analog format and includes huge amount of data, a direct computation using conventional digital methods requires very large data storage devices and very fast computational power. In the conventional approach, the signal is received by a sensing device and transformed into digital signals by an analog-to-digital converter. The huge amount of converted digital signals are condensed by the next digital processing stage. On the other hand, neural sensory circuits can combine the sensing, conversion, and feature extraction processes into a single process. The sensory systems of various animals convert the real-world signals into the internal representation for neurocomputing in the brain, which include information localization and early mapping process.

The neural circuit of a sensory system differs from that of the brain in its highly localized neural structure. One neuron in the sensory system is tied to just the neighboring neurons. Thus, the information in the neural sensory system is locally processed and gradually spread to the whole network in a decay format. The high locality property is seldom explored in other neural networks. Due to this locality property, the interconnection problem in VLSI implementation can be solved.

One early activity in designing VLSI neural sensory circuits [17] was conducted in Caltech. The entire sensory function was integrated into a sigle chip, while other neural circuits for data processors include only the parallel computing function. Subthrehold operation of an MOS transistor was used not only to save power dissipation but also to utilize the exponential characteristics as shown in Fig. 6.1. The drain current expression for an MOS transistor biased in the subthreshold region is

$$I_{DS} = \mu C_{ox} \frac{W}{L} I_s e^{V_{GS}/nV_T} (1 - e^{-V_{DS}/V_T}),$$ (6.1)

where n is the subthreshold-slope coefficient and I_s is a saturation current similar to a bipolar transistor [18]. Here, V_{GS} is in the range of

$$V_L \leq V_{GS} \leq V_M,$$ (6.2)

where

$$V_L = V_{FB} + \Phi_F + \gamma\sqrt{\Phi_F + V_{SB}},$$ (6.3)

and

$$V_M = V_{FB} + 2\Phi_F + \gamma\sqrt{2\Phi_F + V_{SB}}.$$ (6.4)

Here, Φ_F is the Fermi potential. Due to the exponential characteristics,

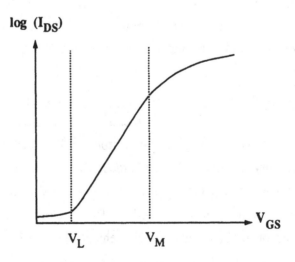

Fig. 6.1 I_{DS} versus V_{GS} of an MOS transistor.

the transfer function of a differential pair stage shown in Fig. 6.2(a) is given as

$$I_{D1} - I_{D2} = I_{ss} \tanh\left[\frac{V_d}{2nV_T}\right] .$$

(6.5)

The transfer function with transistors operating in the strong-inversion region is

$$I_{D1} - I_{D2} = \frac{\mu C_{ox}}{2} \frac{W}{L} V_d \sqrt{\frac{4I_{ss}}{\mu C_{ox} W/L} - V_d^2} ,$$

(6.6)

if the differential-mode input voltage V_d satisfies

$$|V_d| \leq \sqrt{\frac{2I_{ss}}{\mu C_{ox} W/L}} \equiv V_{lmt} .$$

(6.7)

Therefore, transconductance gain G_m of the differential pair at a small differential input voltage is

$$G_m = \sqrt{\mu C_{ox} \frac{W}{L} I_{ss}}$$

(6.8)

in the subthreshold operation and

$$G_m = \frac{I_{ss}}{2nV_T}$$

(6.9)

in the strong-inversion region. Notice that the input dynamic range of the subthreshold operation is limited to approximately nV_T, while that of the strong-inversion operation is limited to V_{lmt}.

By using the exponential characteristics of the subthreshold operation, the analog multiplier and synthesized conductance can be easily constructed as shown in Fig. 6.2(b) and (c). When V_a and V_b of the analog multiplier are very small, output current ΔI becomes

$$\Delta I = K \times V_a V_b .$$

(6.10)

Here, K is conversion coefficient. The analog multiplier is heavily used
in the optical motion detector chip [19] as a local processing block. The
synthesized conductance is composed of the buffer configuration of an
MOS transconductance amplifier. The equivalent conductance value is
the same as the transconductance gain of the differential pair M_1 and M_2
which is given in (6.8) and (6.9). Notice that the conductance is unilateral and the linearity of G_m is limited by the input dynamic range of
the amplifier.

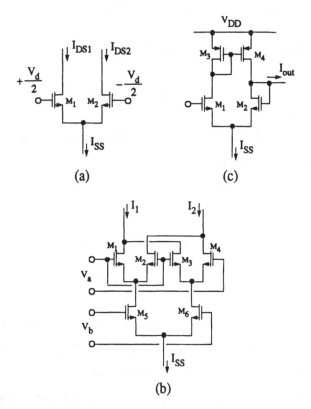

Fig. 6.2 Functional circuits.
(a) Differential pair.
(b) Analog multiplier
(c) Equivalent conductance.

The silicon retina chip [20] performs the sensory functions with preliminary signal processing similar to human organs. Figure 6.3(a) shows the block diagram of one cell in the silicon retina chip. Each cell consists of a photo-receptor and a local processor. The photo-receptor is made from the parasitic PNP bipolar transistor in an n-well CMOS technology. When the pn-junction between the n-diffusion and p-substrate is exposed to light, extra electrons and holes are generated

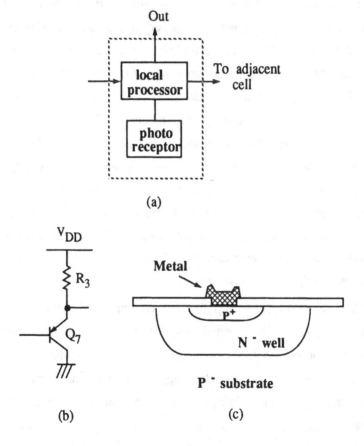

(a)

(b)   (c)

Fig. 6.3 One neuron cell.
    (a) Block diagram.
    (b) Equivalent circuit of photo-receptor
    (c) Cross-section of the PNP transistor.

and appear as a conductive current. In effect, the photo-receptor voltage is modulated by the incoming light intensity. The local processor delivers the light signal to adjacent 6 cells, which is used as an average function to extract the edge of an image. Notice that the silicon retina chip has the hexagonal structure of the cells. Several other photo-receptor structures can also be used in standard CMOS technologies [21].

Two other sensory chips including seehear and optical motion sensor use a similar structure to the silicon retina. In the seehear chip [17], the local processor is just the synthesized conductance as shown in Fig. 6.2(c). Since the conductance is unilateral, the light signals generated by the photo-receptor propagate to only one direction. The strings of the unilateral cells are laid out in the alternative way that the propagating directions of adjacent strings is just reversed. Thus, the light signals exposed at a certain location reach to the ending sides of the seehear chip with some time difference. The difference is converted to the sound signals. Therefore, the sound signals correspond to the angle of an object with respect to the principal direction of the seehear chip. The moving velocity of an object is generated in the optical motion sensor chip whose local processor consists of the synthesized conductance and the analog multiplier as shown in Fig. 6.2.

The silicon cochlea chip [22] is composed of delay elements as shown in Fig. 6.4. Since sounds include only one-dimensional information in the contrast to the video signals for retina, seehear, and motion sensor chips, there is no big challenge for routing the local processors. Several taps after some delays generate a similar response to human cochlea. Here, the delay element was realized with the second-order filter.

Due to poor device characteristics in the subthreshold operation, the neural sensory circuits have non-uniform characteristics. The large offset voltage caused by large device mismatches and temperature coefficient makes poorer image quality from the retina chip than conventional sensory circuits. If the learning capability is incorporated in the neural sensory chips, the non-uniform characteristics can be compensated. The adaptive silicon retina chip fabricated by a double-polysilicon CMOS technology can provide better image uniformity by storing the offset information in the floating-gate transistors [23].

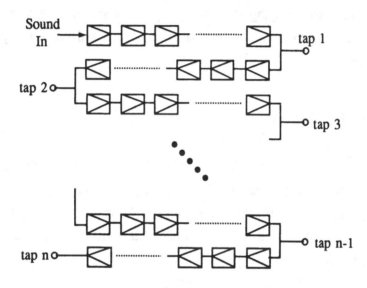

Fig. 6.4 Block diagram of the cochlea chip [22].

6.2 Various Analog Neural Chips

Basic elements of the analog VLSI neural networks consist of operational amplifiers as electronic neurons and resistors as electronic synapses. The neuron transfer function for the software computation is usually an exact mathematical function, while that for the analog implementation can just be an approximated function. Since the neuron transfer function is an important factor to decide the network characteristics, some neural networks, such as the Boltzmann Machine and back-propagation [24,25], can only be approximated in analog VLSI design.

For the VLSI implementation of neural systems, analog approach seems to be extremely attractive in terms of hardware size, power, and speed [26]. In addition, hardware annealing can be easily applied to the VLSI design by changing the amplifier gain. The annealing process is used to find better solution for the pattern matching and scientific optimization applications.

6.2.1 Analog Neurons

Several different circuit topologies for amplifiers in the analog implementation have been used. A moderate voltage gain can be obtained from a CMOS inverter circuit as shown in Fig. 6.5(a). The maximum voltage gain is achieved when both transistors operate in the saturation region,

$$A_v = -\frac{g_{mn} + g_{mp}}{g_{on} + g_{op}} = -\frac{\sqrt{k_n} + \sqrt{k_p}}{\lambda_n + \lambda_p}\frac{1}{\sqrt{I_{Dm}}} , \qquad (6.11)$$

where $k \equiv \mu C_{ox}(W/L)$ and I_{Dm} is the tranasistor current at the maximum voltage gain. The voltage gain for an inverter circuit is in the range of 10 to 100. To achieve a higher voltage gain, a cascaded inverter structure can be used. Due to the device mismatches, the input voltage V_i which gives the maximum voltage gain is widely distributed. The offset voltage can be canceled out with a special switched technique [27].

The circuit schematic for a single-stage CMOS amplifier consisting of the differential-pair input stage and the CMOS inverter-type gain stage is shown in Fig. 6.5(b). With the use of the differential pair, the input offset is reduced. The amplifier low-frequency gain A_{vo} and unity-gain frequency f_u are given as

$$A_{vo} = G_{mi} R_o \tag{6.12}$$

and

$$f_u = \frac{G_{mi}}{C_L} . \tag{6.13}$$

Here, G_{mi} is the transconductance of the differential pair, R_o is the output resistance, and C_L is the load capacitance.

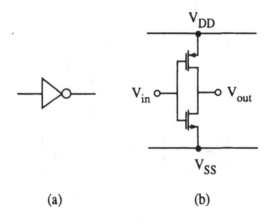

(a)   (b)

Fig. 6.5 A CMOS inverter.
(a) Electrical symbol.
(b) Circuit schematic.

The circuit schematic for a simple 2-stage CMOS amplifier is shown in Fig. 6.6. The 1-st stage consists of the differential pair and current mirror, while the 2-nd stage consists of the driver transistor M_5 and current source M_6. The amplifier low-frequency gain A_{vo} is obtained from the gains of the two stages,

$$A_{vo} = G_{mi} R_1 g_{m5} R_2 , \tag{6.14}$$

where $R_1 = 1/(g_{o2} + g_{o4})$ and $R_2 = 1/(g_{o5} + g_{o6})$, and g_{m5} is the transconductance of M_5. The unity frequency f_u is given as

$$f_u = \frac{1}{2\pi} \cdot \frac{G_{mi}}{C_c} \ . \tag{6.15}$$

To achieve good amplifier stability, the nulling resistor R_z is usually set to be $1/g_{m5}$ [28].

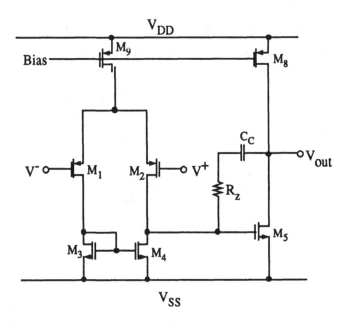

Fig. 6.6 Circuit schematic of a 2-stage CMOS amplifier.

In addition to the amplification function, some arithmetic functions such as addition and subtraction are often required in the analog VLSI neural circuits. The current mirror used for biasing circuitry and current arithmetic is shown in Fig. 6.7(a). Since the transistors operating in the strong inversion are biased with the same gate voltage, the drain current ratio is given as

$$\frac{I_{D2}}{I_{D1}} = \frac{(W/L)_2}{(W/L)_1} \frac{1 + \lambda_2 V_{DS2}}{1 + \lambda_1 V_{DS1}} \ . \tag{6.16}$$

Current addition and subtraction [17] can be conducted with the combi-
nation of current mirrors as shown in Fig. 6.7(b) and (c).

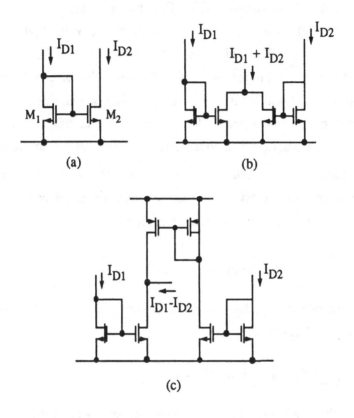

Fig. 6.7 Simple arithmetic cells.
(a) Scaler.
(b) Adder.
(c) Subtractor.

6.2.2 Synapses With Fixed Weights

Due to large input impedance of an MOS amplifier, the high resistivity material should be used for the synapse realization. The high
resistivity material in the standard CMOS technology is well diffusion.

However, the well resistivity is not high enough compared with the output impedance of an MOS amplifier so that a special impedance-matching circuitry is required [7]. One neural chip from AT&T Bell Laboratories [29] used a special process to realize very high resistivity with amorphous-Si. The chip has 256 neurons and 100,000 synapses pre-coded during the fabrication stage. The total chip size is 5700 μm x 5700 μm in a 2.5-μm CMOS technology. The weightings of synapses are defined by electron beams. The main purpose of this chip is known to compress the bandwidth of video image for telephone transmission.

Another way to obtain high resistance is to use the channel resistance from an MOS transistor. When an MOS transistor operates in the triode region, the drain current can be expressed as

$$I_{DS} = \mu C_{ox} \frac{W}{L} \left[(V_{GS} - V_{th})V_{DS} - \frac{V_{DS}^2}{2} \right] \qquad (6.17)$$

if the channel-length modulation effect is neglected. When $V_{DS} \approx 0$, the equivalent conductance is

$$\frac{I_{DS}}{V_{DS}} \approx \mu C_{ox} \frac{W}{L} (V_{GS} - V_{th}) . \qquad (6.18)$$

However, the condition that $V_{DS} \approx 0$ can not be always valid for a practical application. Figure 6.8 shows that the nonlinear conductance of a single MOS transistor can be used at special applications when only inhibitory and excitatory synapses are required [5]. The on/off switches, which are controlled by data stored in the on-chip memory, determine the fixed synapse weight for a content addressable memory. The chip has 54 neurons and 2916 synapses and occupies 6700 μm x 6700 μm area in a 2.5-μm CMOS technology.

A simple way to reduce the nonlinearity is to use both an n-channel transistor and a p-channel transistor as shown in Fig. 6.9. By assuming that $V_1 > V_2$ and the transistors operate in the triode region,

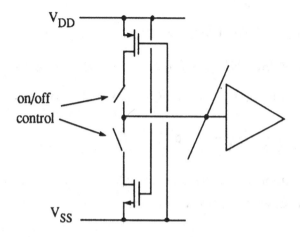

Fig. 6.8 Circuit schematic of a fixed synapse weighting
with ON/OFF control.

the drain currents are

$$I_{DSn} = \mu_n C_{ox} \left[\frac{W}{L}\right]_n \left[(V_{c1}-V_2-V_{thn})(V_1-V_2) - \frac{(V_1-V_2)^2}{2}\right] \quad (6.19)$$

and

$$I_{DSp} = \mu_p C_{ox} \left[\frac{W}{L}\right]_p \left[(V_1-V_{c2}+V_{thp})(V_1-V_2) - \frac{(V_1-V_2)^2}{2}\right]. \quad (6.20)$$

Here, I_{DSn} and I_{DSp} are the drain currents of the n-channel and p-channel transistors, respectively. With the assumption of perfect matches between the n-channel transistor and the p-channel transistor, the design and analysis can be greatly simplified as

$$V_{thn} = -V_{thp} , \quad (6.21)$$

$$\mu_n C_{oxn} (W/L)_n = \mu_p C_{ox} (W/L)_p \; (\equiv \beta) , \quad (6.22)$$

and

$$V_{c1} = -V_{c2} . \tag{6.23}$$

The summation of the currents becomes

$$I_{tot} \equiv I_{DSn} + I_{DSp} = 2\,\beta\,(V_1 - V_2)\,(V_{c1} - V_{thn}) . \tag{6.24}$$

Thus, the conductance T_{eq} of the CMOS switch becomes

$$T_{eq} \equiv \frac{I_{tot}}{V_1 - V_2} = 2\,\beta\,(V_{c1} - V_{thn}) . \tag{6.25}$$

The ideal linear conductance is hardly obtained due to the actual mismatches between the two transistors.

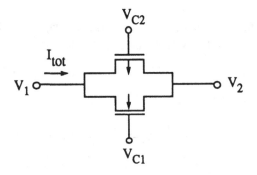

Fig. 6.9 Circuit schematic of a CMOS transmission gate.

The synthesized resistance can be obtained with the circuits originally developed for continuous-time filters [30]. Figure 6.10 shows one such circuitry to eliminate the nonlinearity of an MOS transistor. The drain currents are

$$I_{DS1} = \mu C_{ox} \frac{W}{L} \left[(V_{G1} - V_{th})(V_1 - V_x) - \frac{(V_1 - V_x)^2}{2} \right] , \tag{6.26}$$

$$I_{DS2} = \mu C_{ox} \frac{W}{L} \left[(V_{G2} - V_{th})(V_1 - V_x) - \frac{(V_1 - V_x)^2}{2} \right] , \tag{6.27}$$

$$I_{DS3} = \mu C_{ox} \frac{W}{L} \left[(V_{G3} - V_{th})(V_2 - V_x) - \frac{(V_2 - V_x)^2}{2} \right], \quad (6.28)$$

and

$$I_{DS4} = \mu C_{ox} \frac{W}{L} \left[(V_{G4} - V_{th})(V_2 - V_x) - \frac{(V_2 - V_x)^2}{2} \right]. \quad (6.29)$$

Here, V_x is a constant voltage. By setting $V_{G1} = V_{G4} = -V_{G2} = -V_{G3}$ ($\equiv V_{sw}$), the equivalent conductance is

$$T_{eq} \equiv \frac{I_1 - I_2}{V_1 - V_2} = 2 \mu C_{ox} \frac{W}{L} V_{sw} . \quad (6.30)$$

By connecting the V_x's to the input terminals of an operational amplifier, the linear conductance can be obtained. Since all transistors in this synthesized conductance are all n-channel transistors, the mismatches are minimized.

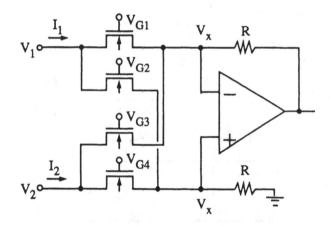

Fig. 6.10 Circuit schematic of a synthesized resistance
and a neuron.

Another approach to use MOS transistors is to operate the transistors in the saturation region as shown in Fig. 6.11. In the current steering circuit structure, the synapse weight is coded into the current level and the current is switched by the differential pair. In this approach, the output current I_{out} is given as

$$I_{out} = \frac{I_S + \Delta I}{2} \, , \tag{6.31}$$

where

$$\Delta I = \begin{cases} \dfrac{\beta}{2}\Delta V_i \sqrt{\dfrac{4I_S}{\beta} - (\Delta V_i)^2} & \text{when } |\Delta V_i| \le V_{lmt} \\ I_S & \text{when } \Delta V_i > V_{lmt} \\ -I_S & \text{when } \Delta V_i < -V_{lmt} \end{cases} \tag{6.32}$$

Here, $\beta = \mu C_{ox}(W/L)$, $\Delta V_i \equiv V_1 - V_{ref}$, and $V_{lmt} = \sqrt{2I_S/\mu C_{ox}(W/L)}$. Since the synapse is nonlinear, this circuit structure can not be used for the synapses of the input layer such as T_{iS} of Hopfield neural-based A/D

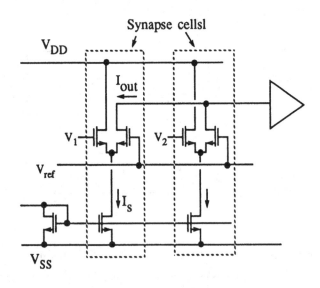

Fig. 6.11 Current steering synapse cell.
          The device aspect ratio of the current
          mirror decides the synapse weight.

converters. However, this synapse can be used for the feedback components of Hopfield networks or for the hidden layers and the output layer. The synapse weights can be coded into device aspect ratios of the current mirror. A programmable synapse of a moderate dynamic range can be made by switching certain ratioed current mirrors [10].

6.2.3 Programmable Synapses

Programmable synapses are used in the network learning process as well as in the compensation of the device mismatches. For real-world applications of the VLSI neural networks, the number of neurons and synapses should be as many as possible. The analog VLSI approach is suitable to realize a compact and electrically programmable synapse. Several commercial neural chips including analog programmable synapses have been developed from Intel Corp., Matushita Laboratory, and AT&T Bell Laboratories.

A fully programmable synapse with EEPROM cells shown in Fig. 6.12 [6] is developed by Intel Corp.. The analog multiplier gives a fairly linear synapse weighting which is decided by the device aspect ratios and the bias current. Since the EEPROM-injected charge at the floating gate can change the threshold voltage of the transistor in the current mirror, bias currents of the analog multiplier are programmable. The output current I_{out} is given as

$$I_{out} = I_{DS1} + I_{DS3} - (I_{DS2} + I_{DS4})$$

$$= \frac{I_{S1} + \Delta I_1}{2} + \frac{I_{S2} - \Delta I_2}{2} - \frac{I_{S1} - \Delta I_1}{2} - \frac{I_{S2} + \Delta I_2}{2}$$

$$= \Delta I_1 - \Delta I_2 . \tag{6.33}$$

Here, ΔI_1 and ΔI_2 can be obtained from (6.32) with $I_S = I_{S1}$ and

$I_S = I_{S2}$, respectively. The equivalent conductance T_{eq} when $\Delta V_i \approx 0$ can be obtained as

$$T_{eq} = \sqrt{\mu C_{ox} \frac{W}{L} I_{S1}} - \sqrt{\mu C_{ox} \frac{W}{L} I_{S2}} \ . \tag{6.34}$$

However, the voltage range V_{IR} for a linear conductance is limited by the smaller dynamic range of the two differential pairs,

$$V_{IR} = \sqrt{\frac{2 \ \min(I_{S1}, I_{S2})}{\mu C_{ox} W / L}} \ . \tag{6.35}$$

Therefore, the effective voltage range is determined by the smaller programming current, while the equivalent conductance is determined by the difference of the two programming currents. For a proper input voltage range and a wide synapse weight range, the device W/L ratio should be very small and the programming currents should be large. Here, the synapse cell occupies a relatively large area of 80 μm x 70 μm in a 1.0-μm CMOS technology.

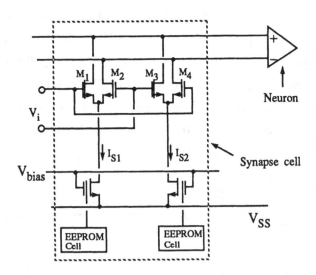

Fig. 6.12 Circuit schematic of a programmable synapse.

The interdependence of the dynamic range and synapse weight range can be broken up by using the BiCMOS cell shown in Fig. 6.13, which is a modified version of the synapse cell in [11]. Transistors M_1 and M_2 convert the synapse input voltage to the currents for diode-connected transistors Q_1 and Q_2. Since the voltages across Q_1 and Q_2 are logarithmic with respect to the currents, the upper differential pairs always operate in the linear region. In this BiCMOS cell, the programming information is stored temporarily at capacitors C_1 and C_2. Due to the isolation requirement between bipolar transistors, the programmable cell size is 80 μm x 80 μm in a 1.0-μm BiCMOS technology.

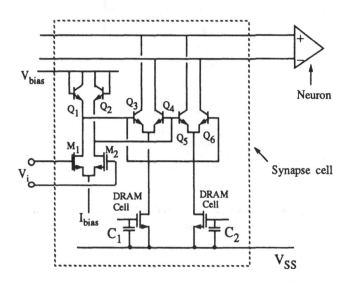

Fig. 6.13   Circuit schematic of a programmable synapse
using BiCMOS process.

Another approach is to use a reconfigurable structure [16] where the synapse accuracy can be adjusted by connecting binary-weighted synapse cells. Each cell consists a 1-bit memory which switches an equivalent conductance. Figure 6.14 shows an example that the synapse weight can be externally arranged for up to a 3-bit resolution. The conductance can be realized by using the channel conductance of an MOS transistor or the current-mode synthesized circuitry which are discussed in the previous sections.

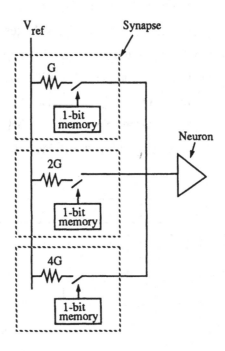

Fig. 6.14   Block diagram of the reconfigurable synapses
and neurons [16].

6.3 Various Digital Neural Chips

There are two different approaches in the digital implementation. The first approach is to implement the synapse and neuron with individual circuitry. The neural computation is distributedly conducted in each cell location. The second approach is to store the synapse weight and neuron transfer function information in registers, and to compute them by a powerful processor. Therefore, the neural computation is performed in common cells such as multiplier, adder, and memory. The neural chip of the first approach usually include the same number of neurocomputing cells as the neuron number, while that of the second approach can include just a few number of shared computing cells. In this section, the first approach will be discussed in detail.

Figure 6.15 shows a digitally programmable synapse. The synapse weight is controlled by the turn-on time of the switch and the RC_S time constant. The synapse weight stored in the memory is converted into the pulse width for the switch, T_{on}. The input voltage is converted to current and summed up at the summing capacitor C_S. With the input voltage range of [0 V, V_{iH}], the equivalent conductance T_{eq} is

$$T_{eq} = \frac{T_{on}}{RC_S} V_{iH} .$$ (6.36)

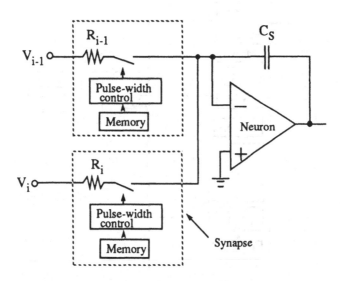

Fig. 6.15 Circuit schematic of a switched-R synapses
and neuron.

After each iteration, the summing capacitor should be reset. The synapse dynamic range and accuracy are totally dependent on the digital circuit. The digital circuitry can be shared using the time-domain multiplexing technique. Due to the periodical integration, the neural circuit operates in the synchronous mode.

The biological neuron-like function which fires according to the neuron threshold is implemented with pulse-stream neurons and synapses [12,31] as shown in Fig. 6.16. In this chip, the neuron consists of an OR-gate, an integrator ($\int$), and a voltage-control oscillator (VCO), while the synapse consists of a pulse width multiplier (PWM). The pulse-streams from synapses are logically summed up and integrated. Thus,

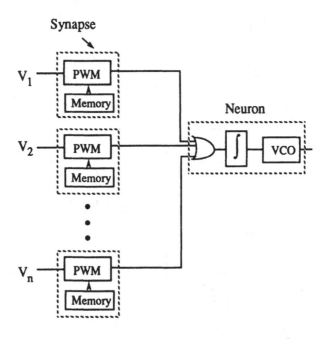

Fig. 6.16 Block diagram of the pulse-stream synapses and
          neuron [12]. A pulse width multiplier is used in each
          synapse cell. The neuron cell consists of an OR-gate,
          an integrator, and a voltage-controlled oscillator.

the integrator output is proportional to the density of summed pulses. Since the VCO output frequency is determined by the integrator voltage, the pulse-stream density of the neuron is dependent on that of the previous neurons and synapse weights.

Another circuit implementation of the pulse-stream approach is to use AND-gates in the neurons [13] as shown in Fig. 6.17. Then, the neuron can be implemented with an AND-gate as shown in Fig. 6.17. The pulse-ON time of each synapse is randomly distributed in the random PWM by the on-chip random-number generator. Due to the randomly distributed pulses, the neuron has a sigmoid-like transfer function. Notice that this stochastism differs in the hardware annealing because the

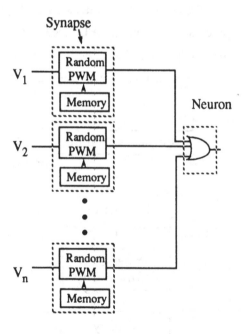

Fig. 6.17   Block diagram of the stochastic synapses and
            neuron [13]. The random number generator pro-
            vides information to the pulse width multiplier of
            each synapse cell.

neuron gain of this approach is always fixed. Since the synapse pulses
are coded in a given time period, the neural chips shown in Fig. 16 and
17 operate synchronously. Thus, the dynamic range of the pulse-width
coded synapses is determined by the ratio of the maximum and the
minimum pulse width. Thus, operating speed is slower as a wide
dynamic range of the synapse weight is larger. In other words, the
neural computing speed of this pulse-width approach is degraded when
the ratio of synapse weights are large.

Fully digital implementations [2,32,33] by using multiprocessors
share the similar computer architecture with conventional digital computers, instead of a direct implementation of the neural network architecture.
The common block diagram of the multiprocessor approach is shown in
Fig. 6.18. Each processor includes a CPU, local memory, and communication circuits. The synapse weights stored in the memory and the neuron outputs from the previous layer are multiplied, summed, and ampli-

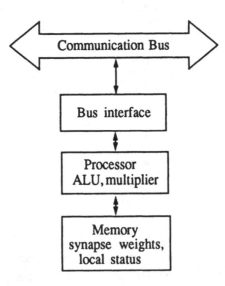

Fig. 6.18   Block diagram of a fully digitally implemented
computing cell.

fied in each processor. The new outputs are transferred to the next processor. The communication channel between the processors is either bus-structure [2,32] or a dedicated connection [33]. Since the general-purpose neural computer has better flexibility than the special-purpose analog VLSI neural chips, a multiprocessor-based neurocomputing device is suitable for accelerating the execution of software algorithms. Recently, this approach is widely commercialized for image recognition and vision processing applications.

Another major category of neural VLSI chips is associative memory [34]. There are two kinds of associative memories: the auto-associative and hetero-associative memories. In the auto-associative memory, the stored content with a partially contained input pattern is the best matched output. On the other hand, the hetero-associative memory produces the output by mapping the input to another pattern [35]. A commercial chip [36] based on this architecture is used for address decoders in computer networking application.

References

[1]   B. Widrow and Bernard, "An adaptive 'Adaline' neuron using chemical 'Memisters'," *Stanford Electronics Lab.,* Technical Report Number 1553-2, Oct. 17, 1960.

[2]   R. H. Nielsen, "Neural-computing: picking the human brain," *IEEE Spectrum,* vol. 25, no. 3, pp. 36-41, Mar. 1988.

[3]   P. Treleaven, M. Pacheco, and M. Vellasco, "VLSI architectures for neural networks," *IEEE Micro Magazine,* vol. 9, no. 6, pp. 8-27, Dec. 1989.

[4]   M. Sililotti, M. R. Emerling, and C. Mead, "VLSI architectures for implementation of neural networks," *Neural Networks for Computing, AIP Conf. Proc. 151,* Editor: J. S. Denker, pp. 408-413, Snowbird, UT, 1986.

[5]   H. P. Graf, L. D. Jackel, and W. E. Hubbard, "VLSI implementation of a neural network model," *IEEE Computer Magazine,* vol. 21, no. 3, pp. 41-49, Mar. 1988.

[6]   M. Holler, S. Tam, H. Castro, R. Benson, "An electrically trainable artificial neural network (ETANN) with 10240 'float gate' synapses," *Proc. of IEEE/INNS Inter. Joint Conf. Neural Networks,* vol. II, pp. 191-196, Washington D.C., June 1989.

[7]   B. W. Lee and B. J. Sheu, "Design of a neural-based A/D converter using modified Hopfield network," *IEEE Jour. of Solid-State Circuits,* vol. SC-24, no. 4, pp. 1129-1135, Aug. 1989.

[8]   K. Goser, U. Hilleringmann, U. Rueckert, and K. Schumacher, "VLSI technologies for artificial neural networks," *IEEE Micro Magazine,* vol. 9, no. 6, pp. 28-44, Dec. 1989.

[9]   R. E. Howard, D. B. Schwartz, J. S. Denker, R. W. Epworth, H. P. Graf, W. E. Hubbard, L. D. Jackel, B. L. Straughn, and D. M. Tennant, "An associative memory based on an electronic neural network architecture," *IEEE Trans. on Electron Devices,* vol. ED-34, no. 7, pp. 1553-1556, July 1987.

[10] P. Mueller, J. V. D. Spiegel, D. Blackman, T. Chiu, T. Clare, C. Donham, T. P. Hsieh, M. Loinaz, "Design and fabrication of VLSI components for a general purpose analog neural computer," in *Analog VLSI Implementation of Neural Systems,* Editors: C. Mead and M. Ismail, Boston, MA: Kluwer Academic, pp. 135-169, 1989.

[11] T. Morishita, Y. Tamura, and T. Otsuki, "A BiCMOS analog neural network with dynamically updated weights," *Tech. Digest of IEEE Inter. Solid-State Circuits Conf.*, pp. 142-143, San Francisco, CA, Feb. 1990.

[12] A. F. Murray, "Pulse arithmetic in VLSI neural network," *IEEE Micro Magazine*, vol. 9, no. 6, pp. 64-74, Dec. 1989.

[13] D. E. Van den Bout and T. K. Miller III, "A digital architecture employing stochasticism for the simulation of Hopfield neural nets," *IEEE Trans. on Circuits and Systems*, vol. 36, no. 5, pp. 732-746, May 1989.

[14] A. Chiang, R. Mountain, J. Reinold, J. LaFranchise, J. Gregory, and G. Lincoln, "A programmable CCD Signal Processor," *Tech. Digest of IEEE Inter. Solid-State Circuits Conf.*, pp. 146-147, San Francisco, CA, Feb. 1990.

[15] C. F. Neugebauer, A. Agranat, and A. Yariv, "Optically configured phototransistor neural networks," *Proc. of IEEE/INNS Inter. Joint Conf. on Neural Networks*, vol. 2, pp. 64-67, Washington D.C., Jan. 1990.

[16] H. P. Graf and D. Henderson, "A reconfigurable CMOS neural networks," *Tech. Digest of IEEE Inter. Solid-State Circuits Conf.*, pp. 144-145, San Francisco, CA, Feb. 1990.

[17] C. A. Mead, *Analog VLSI and Neural Systems*, New York: Addison-Wesley, 1989.

[18] Y. P. Tsividis, *Operation and Modeling of the MOS Transistor*, New York: McGrow-Hill, 1987.

[19] J. E. Tanner and C. A. Mead, "A correlating optical motion detector," *Proc. of Conf. on Advanced Research in VLSI*, Dedham, MA: Artech House, 1984.

[20] M. A. Sivilotti, M. A. Mahowald, and C. A. Mead, "Real-time visual computations using analog CMOS processing arrays," *Proc. of the Stanford Advanced Research in VLSI Conference*, Cambridge, MA: The MIT Press, 1987.

[21] K. Goser, U. Hilleringmann, U. Rueckert, and K. Schumacher, "VLSI technologies for artificial neural networks," *IEEE Micro Magazine*, vol. 9, no. 6, pp. 28-44, Oct. 1989.

[22] J. Lazzaro and C. Mead, "Circuit models of sensory transduction in the cochlea," in *Analog VLSI Implementation of Neural Systems*, Editors: C. Mead and M. Ismail, Boston, MA: Kluwer Academic, pp.85-102, 1989.

[23] C. Mead, "Adaptive retina," in *Analog VLSI Implementation of Neural Systems*, Editors: C. Mead and M. Ismail, Boston, MA: Kluwer Academic, 1989.

[24] D. E. Rumelhart and J. L. McClelland, *Parallel distributed processing, vol. 1: Foundations*, Chapter 7, Cambridge, MA: The MIT Press, 1987.

[25] W. P. Jones and J. Hoskins, "Back-propagation: a generalized delta learning rule," *Byte Magazine*, pp. 155-162, Oct. 1987.

[26] Y. P. Tsividis, "Analog MOS integrated circuits - certain new ideas, trends, and obstacles," *IEEE Jour. Solid-State Circuits*, vol. SC-22, no. 3, pp. 351-321, June, 1987.

[27] Y. S. Yee, L. M. Terman, and L. G. Heller, "A 1mV MOS comparator," *IEEE Jour. of Solid-State Circuits*, vol. SC-13, no. 3, pp. 294-298, June 1978.

[28] P. R. Gray and R. G. Meyer, *Analysis and Design of Analog Integrated Circuits*, 2nd Ed., New York: John Wiley & Sons, 1984.

[29] H. P. Graf, L. D. Jackel, R. E. Howard, B. Straughn, J. S. Denker, W. Hubbard, D. M. Tennant, and D. Schwartz, "VLSI implementation of a neural network memory with several hundreds of neurons," *Neural Networks for Computing, AIP Conf. Proc. 151*, Editor: J. S. Denker, pp. 182-187, Snowbird, UT, 1986.

[30] M. Ismail, S. V. Smith, and R. G. Beale, "A new MOSFET-C universal filter structure for VLSI," *IEEE Jour. of Solid-State Circuits*, vol. SC-23, no. pp. 183-194, Feb. 1988.

[31] A. Agrapat, A. Yariv, "A new architecture for a microelectronic implementation of neural network models," *Proc. of IEEE First Inter. Conf. on Neural Networks*, vol. III, pp. 403-409, San Diego, CA, June 1987.

[32] D. Hammestrom, "A VLSI architecture for high-performance, low-cost, on-chip learning," *Proc. of IEEE/INNS Inter. Conf. on Neural Networks*, vol. II, pp. 537-544, San Diego, CA, June 1990.

[33] H. Kato, H. Yoshizawa, H. Iciki, and K. Asakawa, "A parallel neuroncomputer architecture towards billion connection updates per second," *Proc. of IEEE/INNS Inter. Joint Conf. on Neural Networks*, vol. II, pp. 47-50, Washington D.C., Jan. 1990.

[34] T. Kohonen, *Self-Organization and Associative Memory*, 2nd Ed., New York: Springer-Verlag, 1987.

[35] " A heteroassociative memory using current-mode MOS analog VLSI circuits," *IEEE Trans. on Circuits and Systems*, vol. 36, no. 5, pp. 747-755, May 1989.

[36] Am99C10 256x48 Content Addressable Memory Datasheet, Advanced Micro Devices Inc., Sunnyvale, CA, Feb. 1989.

Chapter 7

Conclusions and Future Work

Due to the massively parallel processing capabilities, VLSI neural systems are critical to the success of future high-performance, low cost computing machines. To design analog neural chips, VLSI architectures have to be developed from the neural network algorithms. In this book, theory and practical design examples for analog VLSI neural networks with hardware annealing have been described. We especially focus on

1. Dynamics of Hopfield neural networks and techniques to eliminate local minima from the energy function;

2. Competitive-Hopfield networks to solve combinatorial optimization with multiple constraint functions;

3. Hardware annealing for fast searching of optimal solution by changing the amplifier voltage gain;

4. Programmable synapses and gain-adjustable neurons for multi-layered VLSI neural systems; and

5. Various analog and digital VLSI neural chips.

Hardware annealing is the key to achieve optimal solutions in a very short period of time because the annealing process is carried out in parallel and the voltage gain of the neurons is changed continuously. The comparison of the self-correction method and hardware annealing method for Hopfield neural networks is listed in Table 7.1.

Table 7.1 Comparison of correction and annealing schemes

	Self-Correction Technique	Hardware Annealing Technique
Computing Speed	Fast	Fast
Device Matching	Accurate	Not accurate
Applicable Network	Only Hopfield net	Any network
Limitation	Global minima should be well-known.	Effective for asynchronous net.

The DRAM-style synapse cells for medium-term weight storage and the EEPROM-style synapse cells for long-term weight storage have been presented. By using these design styles, several prototyping neural chips operating in either asynchronous or synchronous modes have been designed, fabricated, and tested. Two practical applications including neural-based A/D conversion and the digital image restoration are discussed in detail. By using the compact synapse and neuron circuit cells, an electrically programmable neural chip containing about 300 neurons and 100k synapses can be fabricated in the state-of-the-art 1-µm CMOS technology.

Further studies to improve the performance and to reduce the area of synapse cells and neuron cells are necessary to integrate a larger neural network into a single VLSI chip. The dynamic range of programmable synapse weights with fixed standby power dissipation can be greatly enhanced by using the nonsaturated input stage architecture described in Appendix B. Since the nonsaturated differential pair always operates in the high-gain region, the power dissipation of reconfigurable

VLSI neural circuits can be reduced dramatically.

Design automation can be utilized to build VLSI neural chips because of functional regularities in neural networks. In general, analog VLSI circuits are sensitive to device mismatches and parasitic components, so that design automation for the analog circuits is still very primitive. However, components in neural networks do not have to be very precise or fast. In addition, the self-learning capability in neural networks can compensate not only initial device mismatches but also long-term drift of the device characteristics. Several design methodologies for analog VLSI circuits, which include analog array, standard cells, parameterized cells, and programmable cells, can be used. Since neural networks have a high degree of modularity consisting of a neuron and associated synapses, the analog standard-cell approach is quite suitable to design VLSI neural circuits.

Advanced development toward system integration for large and practical problems can be done by using the programmable synapses and the gain-adjustable neurons. Research on on-chip learning is also very important to reduce total system learning time and to compensate nonideal characteristics of analog VLSI neural chips. Major design factors in analog VLSI neural circuits include device mismatches, limited synapse weights, and imprecise neuron transfer functions. Thus, a dedicated on-chip learning circuitry, which monitors the network outputs and adjusts the synapse weights, should be developed for real-world applications.

Appendixes

Appendix A

Program for Neural-Based A/D Conversion and Traveling Salesman Problems

Many subroutines are commonly used for the two problems. To compile, just type " cc Neural_ADC.c -lm " or " cc TSP.c -lm ".

```
/************** filename = Neural_ADC.c ********************
******************* main   program *********************
*********************************************************/

#include "declarat.c"

main()
{
  int  i, int_iter=0, initial_cond_flag, E_eq =0;
  double E_old = 1.0e30, E_new, temp,
      U_old[N_bit], U[N_bit], V_old[N_bit], V[N_bit],
      U2[N_bit], U1H[N_bit], V2[N_bit], V1H[N_bit],
      find_E(), est_local_error(),
      Time_final =1.0, factor_etol,
      local_error;            /* local error tolerance */

  ask_initial_values();            /* get parameters */
```

```
printf(" Analog Input voltage Vs = ");
scanf("%lf", &VS);
printf(" Vs = %f, Vref = %f ",VS,VR);

printf("Set initial condition at input/0 or at output/1");
                                    /* set initial conditions */
while ( scanf("%d", &initial_cond_flag ) )
  if ( initial_cond_flag == 0 | initial_cond_flag == 1 ) break;
if ( initial_cond_flag == 0 )
  {
    printf(" Set input initial conditions ");
      init_soln_o(U_old);
      for ( i = 0; i < N_bit ; ++i )
        {
          U_old[i] -= 0.5;
          U_old[i] /= 10.0;
          U[i] = U_old[i];
        }
    print_U(U);
    find_V(U,V_old);
  }
else
  {
    printf(" Set output initial conditions ");
    init_soln_o(V_old);
    for ( i=0; i < N_bit ; ++i ) U_old [i] =0.0;
  }
for ( i = 0; i < N_bit; i++) V[i] = V_old[i];
print_V(V);

cre_I();
cre_T();
cre_Tc();
while ( iterate == 'y')
  {
                                    /* iteration start */
```

```
++ iter_count;
            /* Runge-kutta 4-th method or Euler forward method
                                        with adaptive step control */
euler_model(U_old,V_old,U2,V2,wu);       /* regular time step */
int_iter = 0;
local_error = 1.0e30;
factor_etol = 1.0;
temp =1.0;

   while ( temp > 0.0 && fix_T != 'y' )
     {
     if ( ++ int_iter > 100 | factor_etol < 1.0e-10 )break;
     wu /= 2.0;
     euler_model(U_old,V_old,U1H,V1H,wu);
                         /* 1st time with half step size      */
     euler_model(U1H,V1H,U,V,wu);
                         /* 2nd time with half step size      */
     local_error = est_local_error(U2,U);
                         /* estimate local error */
     E_new = find_E(0,V);
     temp =local_error - etol/(Time_final*1.0);
     if (batch ==0 )
       {
       printf("interal iter= %d, energy= %g,",++int_iter, E_new);
       printf(" local error= %e, wu= %g ",local_error,wu);
       }
     for (i=0; i < N_bit; ++i)
       {
          U2[i] = U1H[i];
          V2[i] = V1H[i];
       }
     }
   if ( fix_T == 'y' )
      for ( i=0; i < N_bit; ++i){U[i] = U2[i]; V[i] = V2[i]; }
   time += wu;
```

```
        printf("iter= %d time= %e ", iter_count, time );

                        /* terminating criteria by E updating */
    E_new = find_E(1,V);
    if ( fabs( E_new - E_old ) < 1.0e-6)
        ++E_eq;
     else
          E_eq = 0;
    if ( E_eq > 10 )terminate_flag =1;
    E_old = E_new;
    if ( batch == 0 l terminate_flag == 1 ) print_V(V);
    if ((iter_count / 100) * 100 - iter_count == 0) print_V(V);
    if ( iter_count > iteration_lmt ) terminate_flag = 1;
    for (i=0; i < N_bit; ++i )
       {
       U_old[i] = U[i];
       V_old[i] = V[i];
       }
    if ( fix_T != 'y')
       { if (local_error < 1.0e-10)   /* next step wu */
           local_error =1.0e-10;
         wu *= pow (etol/(local_error*Time_final), 1.0 );
         if ( wu > 1.0e-1 )wu=1.0e-1;
       }
    if ( terminate_flag == 1)break;
  }
  output(U,V);
}                               /* end of  main program */



/****************** filename = tsp.c **********************
****************** main   program *********************
**********************************************************/
```

```c
#include "declarat.c"

main()
{
  char ch_const, iterate, fix_T;
  int  i, int_iter=0, initial_cond_flag, E_eq =0, itemp;
  double E_old = 1.0e30, E_new, temp,
       U_old[N_SQ], U[N_SQ], V_old[N_SQ], V[N_SQ],
       U2[N_SQ], U1H[N_SQ], V2[N_SQ], V1H[N_SQ],
       TD[N_SQ][N_SQ]
       find_E(), est_local_error(),
       Time_final =1.0, factor_etol,
       local_error;              /* local error tolerance */

  printf("Type seed for distance matrix !!");
  if ( N == 10 ) printf("0 = fixed location");
  scanf("%d", &itemp);
  if ( itemp = 0 )
    init_dist();
  else
    {
      for ( i = 0; i < itemp; ++i ) random_city_coord(0);
      random_city_coord(1);
    }
  printf(" seed = %d ", itemp);

  cre_delta();
  set_up_dist();

  ask_initial_values();                  /* get parameters */

  printf("Set initial condition at input/0 or at output/1");
                                         /* set initial conditions */
  while ( scanf("%d", &initial_cond_flag ) )
    if ( initial_cond_flag == 0 | initial_cond_flag == 1 ) break;
  if ( initial_cond_flag == 0 )
    {
```

```
   printf(" Set input initial conditions ");
      init_soln_o(U_old);
      for ( i = 0; i < N_SQ ; ++i )
        {
         U_old[i] -= 0.5;
         U_old[i] /= 10.0;
         U[i] = U_old[i];
        }
     print_U(U);
     find_V(U, V_old, initial_cond_flag);
  }
 else
  {
     printf(" Set output initial conditions ");
     init_soln_o(V_old);
     for ( i=0; i < N_SQ ; ++i ) U_old [i] =0.0;
  }
for ( i = 0; i < N_SQ; i++) V[i] = V_old[i];
print_V(V);

cre_I();
cre_T(TD);
cre_Tc(TD);

printf(" Version 1, May 10 1989 ");
printf("Euler w/ adaptive, w/o symmetric solutions");
iterate = 'y';
printf("Batch ? Yes/1");
scanf("%d", &batch);

printf("Number of iterations : ");
scanf("%d", &iteration);

printf("Fixed Time Step ? y or n");
while ( fix_T = getchar() )
  if ( fix_T == 'y' | fix_T == 'n') break;
if ( fix_T == 'y') printf(" fixed time step mode ");
```

```
else printf(" adaptive time step mode ");

printf("Use competative network ? yes or no");
while ( comp_net = getchar() )
  if ( comp_net == 'y' | comp_net == 'n' ) break;
if ( comp_net == 'y' )
  {
    printf(" Use Competative network ");
    printf("Adaptive threshold voltage for competative net ?
          y or n");
    while ( adapt_th = getchat() )
      if ( adapt_th == 'y' | adapt_th == 'n' ) break;
    if ( adapt_th == 'y') printf(" Adaptive threshold mode ");
  }
else
  printf(" Use Hopfield network Only ");
while ( iterate == 'y')
```

```
{
                                          /* iteration start */
++ iter_count;
    /*** Runge-kutta 4-th method or Euler forward method
                              with adaptive step control */
euler_model(U_old,V_old,U2,V2,wu);   /* regular time step */
int_iter = 0;
local_error = 1.0e30;
factor_etol = 1.0;
temp =1.0;

while ( temp > 0.0 && fix_T != 'y' )
  {
    if ( ++ int_iter > 100 I factor_etol < 1.0e-10 ) break;
    wu /= 2.0;
    euler_model(U_old,V_old,U1H,V1H,wu);
                              /* 1st time with half step size */
    euler_model(U1H,V1H,U,V,wu);
                              /* 2nd time with half step size */
    local_error = est_local_error(U2,U);
                              /* estimate local error */
    E_new = find_E(0,V);
    temp =local_error - etol/(Time_final*1.0);
    if (batch == 0 )
      {
        printf("interal iter= %d, energy= %g,",++int_iter,
              E_new);
        printf(" local error= %e, wu= %g ",local_error,wu);
      }
    for (i=0; i < N_bit; ++i)
      {
        U2[i] = U1H[i];
        V2[i] = V1H[i];
      }
  }
```

```
    if ( fix_T == 'y' )
      {
        for ( i=0; i < N_bit; ++i) { U[i] = U2[i]; V[i] = V2[i]; }
      }
    if ( comp_net == 'y' )
        del_sym_inv(U, V, 1);
    else
        del_sym_sol(U, V, 1);
      time += wu;
    printf("iter= %d time= %e ", iter_count, time );
                          /* printf("After");   print_V(V); */
                          /* terminating criteria by E updating */
    if ( fabs( E_new - E_old ) < 1.0e-6 )
        ++E_eq;
    else
        E_eq = 0;
    if ( E_eq > 10 ) terminate_flag =1;
    E_old = find_E(1, V);
    if ( batch == 0 | terminate_flag == 1 ) print_V(V);
    if ((iter_count / 100) * 100 - iter_count == 0) print_V(V);
    for (i=0; i < N_SQ; ++i )
      {
        U_old[i] = U[i];
        V_old[i] = V[i];
      }
    if ( fix_T != 'y')
      {
        if (local_error < 1.0e-10)                /* next step wu */
          local_error =1.0e-10;
        wu *= pow( etol/(local_error*Time_final), 1.0 );
        if ( wu > 1.0e-31 ) wu=1.0e-3;
      }
    if ( terminate_flag == 1) break;
  }
output(U,V);
```

```
        printf(" Would you like more iterations ? :"),
        while ( iterate = getchar() )
        if ( iterate == 'y' | iterate == 'n' ) break;
        E_eq = 0;
}                                       /* end of  main program */
```

```
/*************** filename = declarat.c *********************
************* declaration of global variables *****************
************** "include" files and #defines *****************
**********************************************************/

#define N_bit  4          /* number of bits */
#include <math.h>         /* math declarations */
#include <stdio.h>         /* standard io declarations */


int    n,                          /* weighting  constants */
       V_in[N_bit], city[N_bit],   /* vector input for distance
                                      calculation and posn of cities */
       batch=0, terminate_flag =0, /* flags */
       iter_count,                 /* iteration count */
       iteration_lmt;              /* maximum iteration limit */

char   iterate = 'y', fix_T;

double Uo,                         /* gain */
       T[N_bit][N_bit], Tc[N_bit], /* hopfield and correction synapse
                                      matrices , rows for city, col
                                      for posn. */

       TS[N_bit], TR[N_bit],       /* input signal synapse, reference
                                      synapse */

       I[N_bit], Ic[N_bit],        /* input and correction currents */
       time =0.0,
       etol,                       /* local error tolerance */
       VR = -1.0, VS,              /* reference voltage, input signal */
```

```
        wu, wc;
#include "initial.c"
#include "setup.c"
#include "update.c"
#include "output.c"
#include "rk.c"
```

```
/***************** filename = initial.c ********************/

/***** ASK_INITIAL_VALUES *****/
/***** prompt user for the constants *****/

ask_initial_values() {
  printf("Amplifier gain, Uo : ");
  scanf("%le", &Uo);

  printf("Initial time step, wu : ");
  scanf("%le", &wu);

  printf("Correction weighting factor, wc : ");
  scanf("%le", &wc);

  printf("Error tolerance, etol : ");
  scanf("%le", &etol);

  printf("Batch ? Yes/1 ");
  scanf("%d", &batch);

  printf("Number of iterations : ");
  scanf("%d", &iteration_lmt);

  printf("Fixed Time Step ? y or n");
  while ( fix_T = getchar() )
    if ( fix_T == 'y' | fix_T == 'n') break;

/***** start to print title & information *****/
```

```
printf(" *************************** ");
printf(" Version 1, Jan. 22 1990 ");
printf(" Amplifier Gain (Uo) = %e",Uo);
if ( wc == 0.0 )
    printf(" No Correction Mode ");
else
    printf(" Correction Coefficient (wc) = %e ",wc);
printf("initial step size (wu) = %e",wu);
if ( fix_T == 'y')
    printf(" fixed time step mode ");
else
  {
    printf(" adaptive time step mode ");
    printf("error tolerance (etol) = %e", etol);
  }
  if ( batch == 1 )
    printf("Batch mode");
else
    printf("Interactive mode");
  printf("Number of iterations (iteration_lmt) = %d",iteration_lmt);
}


/***** INIT_SOLN_O *****/
/***** initializes the output vector to prespecified values ****/

init_soln_o(V)
double V[N_bit];
{
  char c = 'A';
  int  i, ii;

  printf("Choose the initial condition sets.");
  printf(" 0 --- reset all conditions ");
  printf(" 1 --- randomly generated condition ");
  while ( c = getchar() )
```

```
    if ( c == '0' | c == '1') break;
  switch ( c )
    {
      case '0':                              /* reset all conditions */
            {
              for ( i=0; i < N_bit; ++i)V[i] = 0.0;  /* initialize */
              break;
            }
      case '1':                             /* randomly generated output */
            {
              printf("Type the seed for random number generator");
              scanf("%d", &i);
              for ( ii = 0; ii < i; ++ii) random_gen(V);
              break;
            }
      default:
      break;
    }
}


/****************************************************************/

random_gen(V)
double V[N_bit];
{
  int  i;

  for (i=0; i < N_bit; ++i)
    {
      V[i] = rand();
      while ( V[i] > 1.0 ) V[i] /= 10.0;
                        /*      printf(" V[%d] = %f;",i,V[i]);
                          if ( (i+1) % 4 == 0 )printf("  "); */
    }
}
```

```
/***************** filename = setup.c *********************
************* Constants matrices such as t, delta **************
************* and Tc are set up in this section ***************
*************************************************************/


/***** CRE_T *****/
/***** creates matrix with connection values *****/

cre_T()
{
  int i, j;
  for ( i=0; i < N_bit; ++i)
    for ( j=0; j < N_bit; ++j)
      {
        if ( i == j )
          T[i][j] = 0.0
        else
          T[i][j] = - pow (2.0 , (double)(i + j ));
      }
}


/*************************************************************/
/***** CRE_TC *****/
/***** connection weights for correction term *****/

cre_Tc()
{
  int i, j;
  char dummy, est_avg_w;
  double double_i, temp;
  if ( wc > 0.0 )
    {
      for (i = 0; i < N_bit; i++ )
        {
          Tc[i] = 0.0;
```

```
        for (j = 0; j < i ; j++ )
        Tc[i] = (0.5)*T[i][j];
      }
    }
}


/***************************************************************/
/***** CRE_I *****/
/***** create vector of input currents *****/

cre_I()
{
  int i;
 p for (i = 0; i < N_bit; i++ )
    {
    TS[i] = pow(2.0 , (double) i);
    TR[i] = pow (2.0 , (double) (2*i -1) );
    I[i] = TS[i] * VS + TR[i] * VR;
            /* printf("TS[%d]=%g, TR[%d]=%g, I[%d]=%g",
              i,TS[i],i,TR[i],i,I[i]); */
    }
}




/***************** filename = update.c **********************
*********** update precedures for input and output **********
******* voltages, input current and correction current ******
***********************************************************/


/***** FIND_U *****/
/***** update the input voltage vector *****/

find_U(U, V)
double U[N_bit], V[N_bit];
```

```
{
  int  i, m1, n1;

  for (m1 = 0; m1 < N_bit ; m1++ )
    {
      U[m1]=0.0;
      for (n1 = 0; n1 < N_bit ; n1++ )
      {          U[m1] += T[m1][n1] * V[n1];          }
      U[m1] += I[m1];          /* adding correction current */
      if ( wc != 0.0 ) U[m1] += wc * Ic[m1];
    }
}


/************************************************************/
/***** FIND_V *****/
/***** updates the output voltage vector *****/


find_V(U,V)
double U[N_bit], V[N_bit];
{
  int n1;
  double term1;

  for (n1 = 0; n1 < N_bit; n1++)
     {
     term1 = U[n1]/Uo ;
     V[n1] = (1.0/2.0)*(1 + (tanh(term1)));
     }
}


/************************************************************/
/***** FIND_IC *****/
/***** calculate correction current vector *****/


find_Ic(V)
double V[N_bit];
```

```
{
  int i, j;
  double gout;

  for (i = 0 ; i < N_bit; i++ )
    {
      g(V[i], &gout);
      Ic[i] = Tc[i] * ( 2.0* gout - 1.0 );
    }
}
```

/***/
```
g(x, out)          /* transfer function of correction logic */
double x, *out;
{
  if (x > 1.0e-18 )
    *out =1.0;
  else
    *out = 1.0e18 * x;
}
```

/***/
```
double find_E(flag_prt,V)
int    flag_prt;
double V[N_bit];
{
  int    x, y, i, j, index1, index2;
  double terma=0.0, termb=0.0, E=0.0;

    for ( i = 0 ; i < N_bit ; i++ ) /* first term of energy function */
      for (j = 0 ; j < N_bit ; j++ )
      terma += T[i][j] * V[i] * V[j];

    for ( i = 0 ; i < N_bit ; i++) /* second term of energy function */
      termb += I[i] * V[i];

  E =- 0.5 * terma - termb;
```

```
    if ( flag_prt == 1) printf(" E = %e ",E);
    return (E);
}




/***************** filename = output.c *********************
****************** printing the results ********************
***********************************************************/

/***** OUTPUT *****/
/***** summarizes the output *****/

output(U,V)
double U[N_bit], V[N_bit];
{
    char  pT, pI, pU, pV, dummy;

    printf("Print T matrix ? : ");
    dummy = getchar();
    pT = getchar();
    if (pT == 'y')
      print_T();
    printf("Print I matrix ? : ");
    dummy = getchar();
    pI = getchar();
    if (pI == 'y')
      print_I();
    printf("Print U vector ? : ");
    dummy = getchar();
    pU = getchar();
    if (pU == 'y')
      print_U(U);
    printf("Print V vector ? : ");
    dummy = getchar();
    pV = getchar();
```

```
  if (pV == 'y')
    print_V(V);
}

/**********************************************************
*********** printing the T vector (connections) **********
*********************************************************/

print_T()
{
  int i,j;
  printf("************** T - matrix ****************");
  for (i = 0; i < N_bit; i++ )
    {
      for (j = 0; j < N_bit; j++ )          printf("%g ", T[i][j]);
      printf("  ");
    }
  printf("*******************************************");
}


/*********************************************************/
print_I()
{
  int i;
  printf("*************** I - matrix ******************");
  for (i = 0; i < N_bit; i++)
    printf("%g ", I[i]);
  printf("*******************************************");
}


/*********************************************************/
print_U(U)
double U[N_bit];
{
  int i;
```

```
  printf(" U - matrix : ");
  for (i = 0; i < N_bit; i++)
    {
      if ( (i%N_bit) == 0 )         printf("   ");
      printf("%.3g ", U[i]);
    }
  printf("*******************************************");
}


/**********************************************************/
print_V(V)
double V[N_bit]; {
  int i;
  printf("V - matrix :");
  for (i = 0; i < N_bit; i++)
    {
      if ( (i%N_bit) == 0)          printf("   ");
      printf("%.3g ", V[i]);
    }
  printf("*******************************************");
}




/****************** filename = rk.c **********************/
/***** Runge-Kutta 4-th method *****/


rk4_model(Up,Vo,U,V,H)
double Up[N_bit], Vo[N_bit], U[N_bit], V[N_bit], H;
  {
    int i;
    double U1[N_bit], U2[N_bit], U3[N_bit], U4[N_bit], UT[N_bit];

    /* first */
    find_Ic(Vo);
    find_U(UT,Vo);  /* calculate next input voltage */
```

```
                    /* UT : output, Vo : input      */
for (i=0; i < N_bit; ++i)
  {
   U1[i] = H * UT[i];
   UT[i] = Up[i] + 0.5 * U1[i];
  }

find_V(UT,V,1);    /* update outputs */
                    /* UT : input, V : output */

/* second */
find_Ic(V);
find_U(UT,V);
for ( i=0; i < N_bit; ++i)
  {
   U2[i] = H * UT[i];
   UT[i] = Up[i] + 0.5 * U2[i];
  }
find_V(UT,V,1);

/* third */
find_Ic(V);
find_U(UT,V);
for ( i=0; i < N_bit; ++i)
  {
   U3[i] = H * UT[i];
   UT[i] = Up[i] + U3[i];
  }
find_V(UT,V,1);

/* fourth */
find_Ic(V);
find_U(UT,V);
for ( i=0; i < N_bit; ++i) U4[i] = H * UT[i];

/*** update ***/
for ( i = 0; i < N_bit; ++i)
  {
```

```
      U[i] = Up[i] + (U1[i] + 2.0*U2[i] + 2.0*U3[i] + U4[i])/6.0;
    }
   find_V(U,V,1);
 }




/*************************************************************/
euler_model(Up,Vo,U,V,H)
double Up[N_bit], Vo[N_bit], U[N_bit], V[N_bit], H;
  {
    int i;
    double UT[N_bit];

    if ( wc != 0.0 ) find_Ic(Vo);
    find_U(UT,Vo);  /* calculate next input voltage */
                    /* UT : output, Vo : input       */
    for (i=0; i < N_bit; ++i)
      {
       U[i] = Up[i] + H * UT[i];
      }

    find_V(U,V);   /* update outputs                */
                   /* UT : input, V : output        */
  }




/*************************************************************/
/******* Estimate local error ( Richardson Extrapolation ) ********/

double est_local_error(V1, V2)
double V1[N_bit], V2[N_bit];
  {
    int i;
    double subi, max = 1.0e-30;

    for ( i= 1; i < N_bit; ++i)
```

```
  {
    subi = fabs ( ( V1[i] - V2[i] ));
    if ( subi > max ) max= subi;
  }
  return (max);
}
```

Appendix B

Non-Saturated Input Stage for Wide-Range Synapse Circuits

The differential pair structure shown in Fig. B.1 has been widely used in the design of operational amplifiers. The usefulness of the differential pair stems from its wide common-mode input range (CMIR) and excellent common-mode rejection ratio (CMRR) [1]. However, the dynamic input range of the differential pair is usually very limited. Since one transistor of the differential pair is cut off when a large differential input signal is applied, the corresponding amplifier response is limited by the input stage behavior.

Figure B.2 shows an improved version of the MOS input structure [2,3]. The second differential pair is moved away from the current mirror and the CMIR becomes very symmetrical. With all transistors in the input stage biased in the saturation region, the governing equations of the non-saturated input differential pair can be derived from Kirchoff's Voltage Law,

$$V_d \equiv V^+ - V^- = V_{GS1} - V_{GS6} + V_{GS4} + V_{GS2} \qquad \text{(B.1)}$$

and

$$V_{GS3} - V_{GS5} = V_{GS4} - V_{GS6} . \qquad \text{(B.2)}$$

Here, V_d is the differential input voltage. If the threshold voltage mismatch and transconductance coefficient mismatch are neglected, V_d can be expressed as

$$V_d = \frac{\sqrt{I_{DS1}} - \sqrt{I_{DS2}} + \sqrt{I_{DS6}} - \sqrt{I_{DS4}}}{\sqrt{K/2}} , \qquad \text{(B.3)}$$

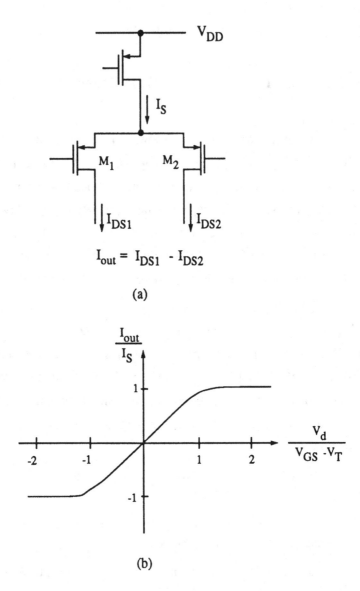

Fig. B.1   Conventional input differential pair in MOS amplifier.
          (a) Circuit schematic.
          (b) Transfer chracteristics.

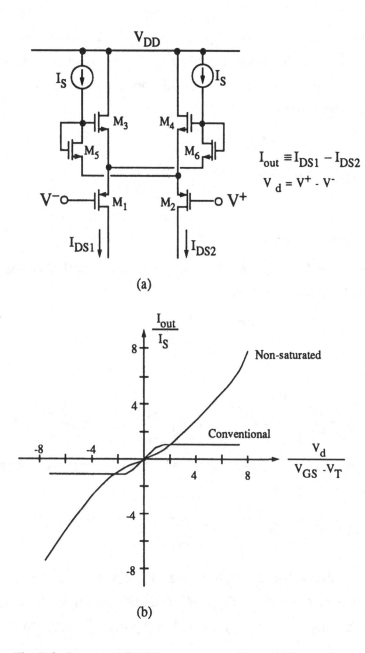

$$I_{out} \equiv I_{DS1} - I_{DS2}$$
$$V_d = V^+ - V^-$$

(a)

(b)

Fig. B.2 Improved CMOS non-saturated input differential
        pair.
        (a) Circuit schematic.
        (b) Transfer characteristics.

where $K = \mu C_{ox} W / L$. Here, device aspect ratios are chosen to achieve the same K value for transistors M_1 to M_6. By using the Kirchhoff's Current Law, the drain currents of MOS transistors are governed by the following relationships,

$$I_{DS1} = I_{DS3} + I_{DS6} , \tag{B.4}$$

$$I_{DS2} = I_{DS4} + I_{DS5} , \tag{B.5}$$

and

$$I_{DS5} = I_{DS6} = I_S . \tag{B.6}$$

Here, I_S is the bias current. A more compact relationship among I_{DS1}, I_{DS2}, and V_d can be obtained,

$$\frac{I_{DS2}}{I_S} = \left[2 - \sqrt{\frac{I_{DS1}}{I_S} - 1} \right]^2 + 1 \tag{B.7}$$

and

$$V_d \sqrt{\frac{K}{2 I_S}} = - \sqrt{\frac{I_{DS1}}{I_S}} - \sqrt{\frac{I_{DS2}}{I_S} - 1} + \sqrt{\frac{I_{DS1}}{I_S} - 1} . \tag{B.8}$$

The ranges of I_{DS1} and I_{DS2} to make all input-stage transistors operate in the saturation region are

$$1 \le \frac{I_{DS1}}{I_S} \le 5 \text{ and } 1 \le \frac{I_{DS2}}{I_S} \le 5 . \tag{B.9}$$

When the differential input voltage V_d is very large, transistor M_4 will be turned off and transistor M_2 will still remain in the saturation region. Currents I_{DS1} and I_{DS2} are decided by the following equations,

$$I_{DS2} = I_S \tag{B.10}$$

and

$$V_d \sqrt{\frac{K}{2 I_S}} = - 2 + \sqrt{\frac{I_{DS1}}{I_S} - 1} + \sqrt{\frac{I_{DS1}}{I_S}} . \tag{B.11}$$

Here, I_{DS1} is larger than $5I_S$. On the other hand, the currents of transistor M_1 and M_2 are decided by the following equations when a large negative V_d is applied,

$$I_{DS1} = I_S \qquad\qquad (B.12)$$

and

$$V_d \sqrt{\frac{K}{2I_S}} = -2 + \sqrt{\frac{I_{DS2}}{I_S} - 1} + \sqrt{\frac{I_{DS2}}{I_S}} . \qquad (B.13)$$

While the drain current of either transistor M_1 or M_2 is limited to the bias current I_S at a large input signal, the drain current difference of the two transistors is unsaturated as shown in Fig. B.2(b). Hence, the non-saturated differential pair gives wider input dynamic range than the conventional differential pair. When the input voltage V_d is very small, the transconductance gain of the non-saturated input stage, G_m, can be determined from (B.8),

$$G_m \equiv \frac{\partial(I_{DS1} - I_{DS2})}{\partial V_d} = \frac{\sqrt{2}}{1 + \sqrt{2}} \sqrt{2I_{DS1}K} . \qquad (B.14)$$

The transconductance gain of the non-saturated input stage is smaller than that of a conventional input stage by a factor 0.59. However, the wider input dynamic range of the non-saturated input stage can make transient response significantly faster.

By using a large bias current, the dynamic range of a programmable synapse using conventional differential-pair stage can be also increased. Since the number of synapses is the square of that of neurons in a fully-connected neural network, total power dissipation is also squarely proportional to the synapse dynamic range. For a large dynamic range, usually very long-channel transistors are used for the differential input stage. Notice that the aspect ratio of these transistors used in Chapter 4 is 3/39. On the other hand, the dynamic range of a synapse using the non-saturated input stage is independent of the bias current. In addition, the synapse cell size can be reduced because transistors with very long channels are not needed in the non-saturated input stage.

References

[1]   P. R. Gray, R. G. Meyer, *Analysis and Design of Analog Integrated Circuits*, 2nd Ed., New York: John Wiley & Sons, 1984.

[2]   B. W. Lee, B. J. Sheu, "A high slew-rate CMOS amplifier for analog signal processing," *IEEE Jour. of Solid-State Circuits*, vol. 25, no. 3, pp. 885-889, June 1990.

[3]   B. W. Lee, B. J. Sheu, "A high-speed CMOS amplifier with dynamic frequency compensation," *Proc. of IEEE Custom Integrated Circuits Conf.*, pp. 8.4.1-8.4.4, Boston, MA, May 1990.

Appendix C

SPICE CMOS LEVEL-2 and LEVEL-4 Model Files

Computer-aided design tools have become indispensable in the integrated-circuit design. The SPICE (Simulation Program with Integrated Circuit Emphasis) program [1] has been widely accepted for circuit analysis since its introduction a decade ago. Circuit simulation execution time has been substantially reduced through algorithm improvement and hardware enhancements in the past few years.

Device modeling plays an important role in VLSI circuit design because computer-aided circuit analysis results are only as accurate as the models used. The SPICE2G6.B and SPICE 3C.1 programs have provided four built-in MOS transistor models [2-5]. The LEVEL-1 model, which contains fairly simple expressions, is most suitable for preliminary analysis. The LEVEL-2 model, which contains expressions from detailed device physics, is widely used for the analysis of analog circuits. Its accuracy is quite limited when small-geometry transistors are used. The LEVEL-3 model represents an early attempt to pursue the semi-empirical modeling approach, which only approximates device physics and relies on the proper choice of the empirical parameters to accurately reproduce device characteristics. The LEVEL-4 model, which is also called BSIM (Berkeley Short-Channel IGFET Model), is ideal for the simulation of digital and analog circuits with effective channel lengths as small as 1 μm.

The modeling equations for the LEVEL-2 model can be found in the books by P. Antognetti and G. Massobrio [6] and by D. Divekar [7]. A set of typical parameter values for the LEVEL-2 model is listed in

Table C.1. These parameter values were extracted from a state-of-the-art 2-μm double-polysilicon p-well CMOS process. The test wafers were fabricated by Orbit Seminconductor Inc. through The MOSIS Service at USC/Information Sciences Institute.

Table C.1. Typical transistor parameters

(From MOSIS 2-μm p-well technology)

parameters	symbols	n-channel	p-channel	unit
V_{tho}	VTO	0.95	-0.81	V
T_{ox}	TOX	3.9 E-8	3.9 E-8	m
X_j	XJ	2.5 E-7	2.5 E-7	m
N_{sub}	NSUB	2.2 E+16	5.8 E+15	cm^{-3}
Φ_f	PHI	0.6	0.6	V
λ	LAMBDA	0.028	0.058	V^{-1}
μ_o	U0	585	273	cm/V-s
μ_{crit}	UCRIT	121762	19133	cm/V-s
μ_{exp}	UEXP	0.221	0.255	--
Υ	GAMMA	0.952	0.518	$V^{+1/2}$
V_{max}	VMAX	72452	36876	m/s
L_d	LD	2.1 E-7	2.4 E-7	m
C_j	CJ	4.06 E-4	2.05 E-4	F/m^2
M_j	MJ	0.46	0.45	--
C_{jsw}	CJSW	4.17 E-10	2.34 E-10	F/m^2
M_{jsw}	MJSW	0.35	0.32	--
C_{GDO}	CGDO	2.69 E-10	3.01 E-10	F/m
C_{GSO}	CGSO	2.69 E-10	3.01 E-10	F/m
C_{GBO}	CGBO	2.00 E-10	2.00 E-10	F/m

The modeling equations for the LEVEL-4 model can be found in [3-5]. The format of the parameters is listed below.

TRANSISTORS

	name	L sens. factor	W sens. factor	units of basic parameter
1	V_{FB} (VFB)	V_{FBl} (LVFB)	V_{FBw} (WVFB)	V
2	ϕ_S (PHI)	ϕ_{Sl} (LPHI)	ϕ_{Sw} (WPHI)	V
3	K_1 (K1)	K_{1l} (LK1)	K_{1w} (WK1)	$V^{1/2}$
4	K_2 (K2)	K_{2l} (LK2)	K_{2w} (WK2)	-
5	η_0 (ETA)	η_{0l} (LETA)	η_{0w} (WETA)	-
6	μ_Z (MUZ)	δ_l (DL)	δ_w (DW)	$cm^2/V{-}s$ (μm, μm)
7	U_{0Z} (U0)	U_{0Zl} (LU0)	U_{0Zw} (WU0)	V^{-1}
8	U_{1Z} (U1)	U_{1Zl} (LU1)	U_{1Zw} (WU1)	$\mu m\, V^{-1}$
9	μ_{ZB} (X2MZ)	μ_{ZBl} (LX2MZ)	μ_{ZBw} (WX2MZ)	$cm^2/V^2{-}s$
10	η_B (X2E)	η_{Bl} (LX2E)	η_{Bw} (WX2E)	V^{-1}
11	η_D (X3E)	η_{Dl} (LX3E)	η_{D_w} (WX3E)	V^{-1}
12	U_{0B} (X2U0)	U_{0Bl} (LX2U0)	U_{0Bw} (WX2U0)	V^{-2}
13	U_{1B} (X2U1)	U_{1Bl} (LX2U1)	U_{1Bw} (WX2U1)	$\mu m\, V^{-2}$
14	μ_S (MUS)	μ_{Sl} (LMS)	μ_{Sw} (WMS)	$cm^2/V^2{-}s$
15	μ_{SB} (X2MS)	μ_{SBl} (LX2MS)	μ_{SBw} (WX2MS)	$cm^2/V^2{-}s$
16	μ_{SD} (X3MS)	μ_{SDl} (LX3MS)	μ_{SDw} (WX3MS)	$cm^2/V^2{-}s$
17	U_{1D} (X3U1)	U_{1Dl} (LX3U1)	U_{1Dw} (WX3U1)	$\mu m\, V^{-2}$
18	T_{ox} (TOX)	T_{emp} (TEMP)	V_{dd} (VDD)	μm (^0C, V)
19	C_{GDO}	CsubGSO	C_{GBO}	F/m
20	XsubPART	DUM1	DUM2	-
21	N_0	N_{0l}	N_{0w}	-

| 22 | N_B | N_{Bl} | N_{Bw} | - |
| 23 | N_D | N_{Dl} | N_{D_w} | - |

INTERCONNECTS

| 1 | R_{sh} (RSH) | C_j (CJ) | C_{jw} (CJW) | I_{js} (IJS) | P_j (PJ) |
| 2 | P_{jw} (PJW) | M_j (MJ) | M_{jw} (MJW) | W_{df} (WDF) | δ_l (DL) |

The names of the process parameters of transistors are listed below:

V_{FB}	flat-band voltage
ϕ_S	surface inversion potential
K_1	body effect coefficient
K_2	drain/source depletion charge sharing coefficient
η_0	zero-bias drain-induced barrier lowering coefficient
μ_Z	zero-bias mobility
U_{0Z}	zero-bias transverse-field mobility degradation coefficient
U_{1Z}	zero-bias velocity saturation coefficient
μ_{ZB}	sensitivity of mobility to the substrate bias at $V_{ds}=0$
η_B	sensitivity of drain-induced barrier lowering effect to the substrate bias
η_D	sensitivity of drain-induced barrier lowering effect to the drain bias, at $V_{ds}= V_{dd}$
U_{0B}	sensitivity of transverse-field mobility degradation effect to the substrate bias
U_{1B}	sensitivity of velocity saturation effect to the substrate bias
μ_S	mobility at zero substrate bias and at $V_{ds}=V_{dd}$

μ_{SB} sensitivity of mobility to the substrate bias at $V_{ds} = V_{dd}$

μ_{SD} sensitivity of mobility to the drain bias at $V_{ds} = V_{dd}$

U_{1D} sensitivity of velocity saturation effect to the drain bias,
at $V_{ds} = V_{dd}$

T_{ox} gate-oxide thickness

T_{emp} temperature at which the process parameters are measured

V_{dd} measurement bias range

N_0 zero-bias subthreshold slope coefficient

N_B sensitivity of subthrehold slope to the substrate bias

N_D sensitivity of subthrehold slope to the drain bias

C_{GDO} gate-drain overlap capacitance per meter channel width

C_{GSO} gate-source overlap capacitance per meter channel width

C_{GBO} gate-bulk overlap capacitance per meter channel length

X_{PART} gate-oxide capacitance model flag

Note: XPART= 0, 0.5, and 1 selects the 40/60, 50/50, and 0/100 channel-charge partitioning methods, respectively.

The names of the process parameters of diffusion layers are listed below:

sheet resistance/square	R_{sh}	Ω/square
zero-bias bulk junction bottom capacitance/unit area	C_j	F/m^2
zero-bias bulk junction sidewall capacitance/unit length	C_{jw}	F/m
bulk junction saturation current/unit area	I_{js}	A/m^2
bulk junction bottom potential	P_j	V
bulk junction sidewall potential	P_{jw}	V

bulk junction bottom grading coefficient M_j -

bulk junction sidewall grading coefficient M_{jw} -

default width of the layer W_{df} m

average reduction of size due to side etching or mask

  compensation δ_s m

The names of the process parameters of poly and metal layers are listed as following:

sheet resistance/square R_{sh} Ω/square

capacitance/unit area C_j F/m^2

edge capacitance/unit length C_{jw} F/m

default width of the layer W_{df} m

average variation of size due to side etching or mask

  compensation δ_l m

The following is an example of a paramter set from The MOSIS Service. The lines starting with "*" are used as comments.

NM1 PM1 DU1 DU2 ML1 ML2
*
*PROCESS=orbit
*RUN=n06m
*WAFER=23
*Gate-oxide thickness= 405.0 angstroms
*Geometries (W-drawn/L-drawn, units are um/um) of transistors

measured were:

* 3.0/2.0, 6.0/2.0, 18.0/2.0, 18.0/5.0, 18.0/25.0
*Bias range to perform the extraction (Vdd)=5 volts
*DATE=08-20-90
*

*NMOS PARAMETERS
*

-8.04245E-01,	6.18948E-02,	-1.30188E-01
7.63757E-01,	0.00000E+00,	0.00000E+00
1.10685E+00,	1.56276E-02,	5.44898E-01
-1.02299E-03,	6.11385E-02,	7.14517E-02
-5.33282E-03,	8.76892E-03,	1.36379E-02
5.57304E+02,	7.36852E-001,	3.19719E-001
5.18361E-02,	3.93568E-02,	-3.35625E-02
1.02034E-02,	6.03913E-01,	-2.73750E-01
1.40810E+01,	-2.87646E+01,	4.41935E+01
-4.24761E-04,	-4.05328E-03,	-1.15756E-03
2.30323E-04,	1.02897E-04,	-5.12080E-03
2.22785E-03,	-1.28163E-02,	2.38689E-02
-2.89587E-03,	1.13662E-03,	1.45730E-02
5.54742E+02,	3.26468E+02,	5.54617E+01
6.25973E+00,	-3.32302E+01,	8.66433E+01
-8.89245E-01,	6.58321E+01,	-1.68777E+01
3.12831E-03,	6.82966E-02,	-2.53679E-02
4.05000E-002,	2.70000E+01,	5.00000E+00
4.71196E-010,	4.71196E-010,	7.36546E-010
1.00000E+000,	0.00000E+000,	0.00000E+000
1.00000E+000,	0.00000E+000,	0.00000E+000
0.00000E+000,	0.00000E+000,	0.00000E+000
0.00000E+000,	0.00000E+000,	0.00000E+000

*

* Gate Oxide Thickness is 405 Angstroms

*

*

*PMOS PARAMETERS

*

-3.56220E-01,	7.02610E-02,	2.87320E-01
6.48989E-01,	4.47941E-24,	-8.06185E-24
5.48415E-01,	-1.58595E-01,	-7.30775E-03
1.75060E-02,	2.25731E-03,	-2.98456E-02
-1.10262E-02,	6.04790E-02,	1.68568E-02
2.37215E+02,	4.30548E-001,	1.14170E-001
1.19621E-01,	4.25473E-02,	-7.87773E-02
1.98410E-02,	3.21916E-01,	-1.02454E-01
9.40135E+00,	-3.73996E+00,	5.98178E+00
-7.32664E-04,	-3.98231E-03,	3.00686E-03
8.70633E-04,	-2.63599E-03,	-4.84968E-03
5.11805E-03,	-1.90571E-03,	3.76107E-03
1.32773E-03,	4.11446E-03,	1.27098E-04
2.38224E+02,	1.63059E+02,	-5.84540E+01
8.43195E+00,	2.38873E+00,	7.93731E+00
-6.08052E-01,	1.28197E+01,	-1.95921E+00
-1.84304E-02,	-4.38722E-03,	1.73589E-02
4.05000E-002,	2.70000E+01,	5.00000E+00
2.75323E-010,	2.75323E-010,	6.86580E-010
1.00000E+000,	0.00000E+000,	0.00000E+000
1.00000E+000,	0.00000E+000,	0.00000E+000
0.00000E+000,	0.00000E+000,	0.00000E+000
0.00000E+000,	0.00000E+000,	0.00000E+000

*

*N+ diffusion::
*

23.5,   4.058400e-04,   4.171400e-10,   0,   0.8
0.8,   0.4644,   0.3512,   0,   0
*

*P+ diffusion::
*

70.89,   2.053400e-04,   2.336600e-10,   0,   0.7
0.7,   0.4431,   0.2451,   0,   0
*

*METAL LAYER -- 1
*

4.700000e-02,   0,   0,   0,   0
0,   0,   0,   0,   0
*

*METAL LAYER -- 2
*

2.500000e-02,   0,   0,   0,   0
0,   0,   0,   0,   0

References

[1]   L. W. Nagel, "SPICE2: A computer program to simulate semicon-
      ductor circuits," *Electron. Res. Lab. Memo ERL-M520,* University
      of California, Berkeley, May 1975.

[2]   A. Vladimirescu, S. Liu, "The simulation of MOS integrated circuits using SPICE2," *Electron. Res. Lab. Memo ERL-M80/7*, University of California, Berkeley, Oct. 1980.

[3]   B. J. Sheu, "MOS transistor modeling and characterization for circuit simulation," *Electron. Res. Lab. Memo ERL-M85/22*, University of California, Berkeley, Oct. 1985.

[4]   B. J. Sheu, D. L. Scharfetter, P. K. Ko, M.-C. Jeng, "BSIM: Berkeley short-channel IGFET model for MOS transistors," *IEEE Jour. of Solid- State Circuits*, vol. SC-22, no. 4, pp. 458-466, Aug. 1987.

[5]   B. J. Sheu, W.-J. Hsu, P. K. Ko, "An MOS transistor charge model for VLSI design," *IEEE Trans. on Computer-Aided Design*, vol. CAD-7, no. 4, pp. 520-527, Apr. 1988.

[6]   P. Antognetti and G. Massobrio, *Semiconductor Device Modeling with SPICE*, New York: McGraw-Hill, 1988.

[7]   D. A. Divekar, *FET Modeling for Circuit Simulation*, Boston, MA: Kluwer Academic, 1988.

Bibliography

[1] 80170NW Electricaly Trainable Analog Neural Network, Intel Corporation, Santa Clara, CA, May 1990.

[2] Am99C10 256x48 Content Addressable Memory Datasheet, Advanced Micro Devices Inc., Sunnyvale, CA, Feb. 1989.

[3] E. H. L. Aarts and P. J. M. van Laarhoven, "A new polynomial-time cooling schedule," *Proc. of IEEE Inter. Conf. on Computer-Aided Design*, pp. 206-208, Nov. 1985.

[4] A. Agrapat, A. Yariv, "A new architecture for a microelectronic implementation of neural network models," *Proc. of IEEE First Inter. Conf. on Neural Networks*, vol. III, pp. 403-409, San Diego, CA, June 1987.

[5] Y. Akiyama, A. Yamashita, M. Kajiura, and H. Aiso, "Combinatorial optimization with Gaussian Machines," *Proc. of IEEE/INNS Inter. Joint Conf. on Neural Networks*, vol. I, pp. 533-540, Washington D.C., June 1989.

[6] J. A. Anderson, "A simple neural network generating an interactive memory," *Mathematical Biosciences*, vol. 14, pp. 197-220, 1972.

[7] J. A. Anderson, E. Rosenfeld, *Neuralcomputing - Foundation of Research*, Cambridge, MA: The MIT Press, 1988.

[8]   H. C. Andrews and B. R. Hunt, *Digital Image Restoration* Englewood Cliffs, NJ: Prentice-Hall Inc., 1977.

[9]   P. Antognetti and G. Massobrio, *Semiconductor Device Modeling with SPICE*, New York: McGraw-Hill, 1988.

[10] K. A. Boahen, P. O. Pouliquen, A. G. Andreou, and R. E. Jenkins, " A heteroassociative memory using current-mode MOS analog VLSI circuits," *IEEE Trans. on Circuits and Systems*, vol. 36, no. 5, pp. 747-755, May 1989.

[11] A. Chiang, R. Mountain, J. Reinold, J. LaFranchise, J. Gregory, and G. Lincoln, "A programmable CCD Signal Processor," *Tech. Digest of IEEE Inter. Solid-State Circuits Conf.*, pp. 146-147, San Francisco, CA, Feb. 1990.

[12] G. Dahlquist and A. Björck, *Numerical Methods*, pp. 330-350, Englewood Cliffs, NJ: Prentice-Hall, 1974.

[13] G. Dahlquist and A. Björck, *Numerical Methods*, pp. 269-273, Englewood Cliffs, NJ: Prentice-Hall, 1974.

[14] J. Dayhoff, *Neural Network Architectures: An Introduction*, New York: Van Nostrand Reinhold, 1990.

[15] D. A. Divekar, *FET Modeling for Circuit Simulation*, Boston, MA: Kluwer Academic, 1988.

[16] R. K. Ellis, "Fowler-Nordheim emission from non-planar surfaces," *IEEE Electron Device Letters*, vol. 3, no. 11, pp. 330-332, Nov. 1982.

[17] W.-C. Fang, B. J. Sheu, and J.-C. Lee, "Real-time computing of optical flow using adaptive VLSI neuroprocessors," *Proc. of IEEE Inter. Conf. on Computer Design'* Cambridge, MA, Sept. 1990.

[18] S. Geman and D. Geman, "Stochastic relaxation, gibbs distributions, and the bayesian restoration of images," *IEEE Trans. on Pattern Analysis and Machine Intelligence*, vol. PAMI-6, no. 6, pp. 721-741, Nov. 1984.

[19] K. Goser, U. Hilleringmann, U. Rueckert, and K. Schumacher, "VLSI technologies for artificial neural networks," *IEEE Micro Magazine*, vol. 9, no. 6, pp. 28-44, Dec. 1989.

[20] S. M. Gowda, B. W. Lee, and B. J. Sheu, "An improved neural network approach to the traveling salesman problem," *Proc. of IEEE Tencon*, pp. 552-555, Bombay, India, Nov., 1989.

[21] H. P. Graf and P. deVegvar, "A CMOS implementation of a neural network model," *Proc. of the Stanford Advanced Research in VLSI Conference*, pp. 351-362, Cambridge, MA: The MIT Press, 1987.

[22] H. P. Graf and D. Henderson, "A reconfigurable CMOS neural networks," *Tech. Digest of IEEE Inter. Solid-State Circuits Conf.*, pp. 144-145, San Francisco, CA, Feb. 1990.

[23] H. P. Graf, L. D. Jackel, R. E. Howard, B. Straughn, J. S. Denker, W. Hubbard, D. M. Tennant, and D. Schwartz, "VLSI implementation of a neural network memory with several hundreds of neurons," *Neural Networks for Computing, AIP Conf. Proc. 151*, Editor: J. S. Denker, pp. 182-187, Snowbird, UT, 1986.

[24] H. P. Graf, L. D. Jackel, and W. E. Hubbard, "VLSI implementation of a neural network model," *IEEE Computer Magazine*, vol. 21, no. 3, pp. 41-49, Mar. 1988.

[25] P. R. Gray and R. G. Meyer, *Analysis and Design of Analog Integrated Circuits*, 2nd Ed., New York: John Wiley & Sons, 1984.

[26] S. Grossberg, *Neural Network and Natural Intelligence*, Cambridge, MA: The MIT Press, 1988.

[27] D. Hammestrom, "A VLSI architecture for high-performance, low-cost, on-chip learning," *Proc. of IEEE/INNS Inter. Conf. on Neural Networks*, vol. II, pp. 537-544, San Diego, CA, June 1990.

[28] D. Hebb, *The Organization of Behavior*, New York: Wiley, 1949.

[29] R. Hecht-Nielsen, "Counter-propagation networks," *Proc. of IEEE First Inter. Conf. on Neural Networks*, vol. II, pp. 19-32, San Diego, CA, 1987.

[30] R. Hecht-Nielsen, "Neural-computing: picking the human brain," *IEEE Spectrum*, vol. 25, no. 3, pp. 36-41, Mar. 1988.

[31] R. Hecht-Nielsen, *Neurocomputing*, New York: Addison-Wesley, 1990.

[32] G. E. Hinton and T. J. Sejnowski, "A learning algorithm for Boltzmann machines," *Cognitive Science*, vol. 9, pp. 147-169, 1985.

[33] M. Holler, S. Tam, H. Castro, R. Benson, "An electrically trainable artificial neural network (ETANN) with 10240 'float gate' synapses," *Proc. of IEEE/INNS Inter. Joint Conf. Neural Networks*, vol. II, pp. 191-196, Washington D.C., June 1989.

[34] P. W. Hollis, J. J. Paulos, "Artificial neural networks using MOS analog multipliers," *IEEE Jour. of Solid-State Circuits*, vol. 25, no. 3, pp. 849-855, June 1990.

[35] J. J. Hopfield, "Neural network and physical systems with emergent collective computational abilities," *Proc. Natl. Acad., Sci. U.S.A.*, vol. 79, pp. 2554-2558, Apr. 1982.

[36] J. J. Hopfield, "Neurons with graded response have collective computational properties like those of two-state neurons," *Proc. Natl. Acad., Sci. U.S.A.*, vol. 81, pp. 3088-3092, May 1984.

[37] J. J. Hopfield and D. W. Tank, " 'Neural' computation of decisions in optimization problems," *Biol. Cybernetics*, vol. 52, pp. 141-152, 1985.

[38] R. E. Howard, D. B. Schwartz, J. S. Denker, R. W. Epworth, H. P. Graf, W. E. Hubbard, L. D. Jackel, B. L. Straughn, and D. M. Tennant, "An associative memory based on an electronic neural network architecture," *IEEE Trans. on Electron Devices*, vol. ED-34, no. 7, pp. 1553-1556, July 1987.

[39] HSPICE Users' Manual H9001, Meta-Software Inc., Campbell, CA, May 1989.

[40] M. Ismail, S. V. Smith, and R. G. Beale, "A new MOSFET-C universal filter structure for VLSI," *IEEE Jour. of Solid-State Circuits*, vol. SC-23, no. pp. 183-194, Feb. 1988.

[41] A. K. Jain, *Fundamentals of Digital Image Processing*, Englewood Cliffs, NJ: Prentice Hall, 1988.

[42] W. P. Jones and J. Hoskins, "Back-propagation: a generalized delta learning rule," *Byte Magazine*, pp. 155-162, Oct. 1987.

[43] H. Kato, H. Yoshizawa, H. Iciki, and K. Asakawa, "A parallel neuroncomputer architecture towards billion connection updates per second," *Proc. of IEEE/INNS Inter. Joint Conf. on Neural Networks*, vol. II, pp. 47-50, Washington D.C., Jan. 1990.

[44] S. Kirkpatrick, C. D. Gelatt, Jr., and M. P. Vecchi, "Optimization by simulated annealing," *Science*, vol. 220, no. 4598, pp. 671-680, May 1983.

[45] A. H. Klopf, "A drive-reinforcement model of single neuron function: an alternative to the Hebbian neural model," *Proc. of Conf. on Neural Networks for Computing*, pp. 265-270, Snowbird UT, Apr. 1986.

[46] T. Kohonen, *Self-Organization and Associative Memory*, 2nd Ed., New York: Springer-Verlag, 1987.

[47] B. Kosko, "Adaptive bidirectional associative memories," *Applied Optics*, vol. 36, pp. 4947-4960, Dec. 1987.

[48] J. Lazzaro and C. A. Mead, "Circuit models of sensory transduction in the cochlea," in *Analog VLSI Implementation of Neural Systems*, Editors: C. A. Mead and M. Ismail, Boston, MA: Kluwer Academic, pp.85-102, 1989.

[49] B. W. Lee and B. J. Sheu, "An investigation on local minima of Hopfield network for optimization circuits," *Proc. of IEEE Inter.*

Conf. on Neural Networks, vol. I, pp. 45-51, San Diego, CA, July 1988.

[50] B. W. Lee and B. J. Sheu, "Design of a neural-based A/D converter using modified Hopfield network," *IEEE Jour. of Solid-State Circuits,* vol. SC-24, no. 4, pp. 1129-1135, Aug. 1989.

[51] B. W. Lee, B. J. Sheu, "A compact and general-purpose neural chip with electrically programmable synapses," *Proc. of IEEE Custom Integrated Circuits Conf.,* pp. 26.6.1-26.6.4, Boston, MA, May 1990.

[52] B. W. Lee, B. J. Sheu, *Design and Analysis of VLSI Neural Networks,* in Neural Networks: Introduction to Theory and Applications, Editor: B. Kosko, Englewood Cliffs, NJ: Prentice-Hall, 1990.

[53] B. W. Lee and B. J. Sheu, "Combinatorial optimization using competitive-Hopfield neural network," *Proc. of IEEE/INNS Inter. Joint Conf. on Neural Networks,* vol. II, pp. 627-630, Washington D.C., Jan. 1990.

[54] B. W. Lee and B. J. Sheu, "Hardware simulated annealing in electronic neural networks," *IEEE Trans. on Circuits and Systems,* to appear in 1990.

[55] B. W. Lee, B. J. Sheu, "A high slew-rate CMOS amplifier for analog signal processing," *IEEE Jour. of Solid-State Circuits,* vol. 25, no. 3, pp. 885-889, June 1990.

[56] B. W. Lee, B. J. Sheu, "A high-speed CMOS amplifier with dynamic frequency compensation," *Proc. of IEEE Custom Integrated Circuits Conf.,* pp. 8.4.1-8.4.4, Boston, MA, May 1990.

[57] B. W. Lee, J.-C. Lee, and B. J. Sheu, "VLSI image processors using analog programmable synapses and neurons," *Proc. of IEEE/INNS Inter. Conf. on Neural Networks,* vol. II, pp. 575-580, San Diego, CA, June 1990.

[58] B. W. Lee, B. J. Sheu, J. Choi, "Programmable VLSI neural chips with hardware annealing for optimal solutions," *IEEE Jour. of*

Solid-State Circuits, to appear.

[59] B. W. Lee, H. Yang, B. J. Sheu, "Analog floating-gate synapses for general-purpose VLSI neural computation," *IEEE Trans. on Circuits and Systems,* to appear.

[60] J.-C. Lee and B. J. Sheu, "Analog VLSI neuroprocessors for early vision processing," in *VLSI Signal Processing IV,* Editor: K. Yao, New York: IEEE Press, Nov. 1990.

[61] J.-C. Lee, B. J. Sheu, "Parallel digital image restoration using adaptive VLSI neural chips" *Proc. of IEEE Inter. Conf. on Computer Design,* Cambridge, MA, Sept. 1990.

[62] R. P. Lippman, "An introduction to computing with neural nets," *IEEE Acoustics, Speech, and Signal Processing Magazine,* pp. 4-22, April 1987.

[63] W. S. McCulloch, W. Pitts, "A logical calculus of the idea immanent in neural nets," *Bulletin of Mathematical Biophysics,* vol. 5, pp. 115-133, 1943.

[64] C. A. Mead, *Analog VLSI and Neural Systems,* New York: Addison-Wesley, 1989.

[65] C. A. Mead, "Adaptive retina," in *Analog VLSI Implementation of Neural Systems,* Editors: C. A. Mead and M. Ismail, Boston, MA: Kluwer Academic, 1989.

[66] T. Morishita, Y. Tamura, and T. Otsuki, "A BiCMOS analog neural network with dynamically updated weights," *Tech. Digest of IEEE Inter. Solid-State Circuits Conf.,* pp. 142-143, San Fransisco, CA, Feb. 1990.

[67] P. Mueller, J. V. D. Spiegel, D. Blackman, T. Chiu, T. Clare, C. Donham, T. P. Hsieh, M. Loinaz, "Design and fabrication of VLSI components for a general purpose analog neural computer," in *Analog VLSI Implementation of Neural Systems,* Editors: C. A. Mead and M. Ismail, Boston, MA: Kluwer Academic, pp. 135-169, 1989.

[68] A. F. Murray, "Pulse arithmetic in VLSI neural network," *IEEE Micro Magazine*, vol. 9, no. 6, pp. 64-74, Dec. 1989.

[69] L. W. Nagel, "SPICE2: A computer program to simulate semiconductor circuits," *Electron. Res. Lab. Memo ERL-M520*, University of California, Berkeley, May 1975.

[70] C. F. Neugebauer, A. Agranat, and A. Yariv, "Optically configured phototransistor neural networks," *Proc. of IEEE/INNS Inter. Joint Conf. on Neural Networks*, vol. 2, pp. 64-67, Washington D.C., Jan. 1990.

[71] R. H. Nielsen, "Neural-computing: picking the human brain," *IEEE Spectrum*, vol. 25, no. 3, pp. 36-41, Mar. 1988.

[72] Y.-H. Pao, *Adaptive Pattern Recognition and Neural Networks*, New York: Addison Wesley, 1989.

[73] D. D. Pollock, *Physical Properties of Materials for Engineers*, Boca Raton, FL: CRC Press, pp. 14-18, 1982.

[74] W. K. Pratt, *Digital Image Processing*, New York: Wiley-Interscience, 1978.

[75] T. Quarles, SPICE3 Version 3C1 Users Guide, *Electron. Res. Lab. Memo UCB/ERL M89/46*, University of California, Berkeley, Apr. 1989.

[76] F. Rosenblatt, *Principles of neurodynamics: perceptrons and the theory of brain mechanisms*, Spartan Books, Washington D.C., 1961.

[77] A. Rosenfeld, A. C. Kak, *Digital Picture Processing*, 2nd Edition, New York: Academic Press, 1982.

[78] D. E. Rumelhart and J. L. McClelland, *Parallel distributed processing, vol. 1: Foundations*, Chapter 7, Cambridge, MA: The MIT Press, 1987.

[79] R. A. Rutenbar, "Simulated Annealing Algorithms: An Overview," *IEEE Circuits and Devices Magazine*, vol. 5, no. 1, pp. 19-26, Jan. 1989.

[80] B. J. Sheu, "MOS transistor modeling and characterization for circuit simulation," *Electron. Res. Lab. Memo ERL-M85/22*, University of California, Berkeley, Oct. 1985.

[81] B. J. Sheu, D. L. Scharfetter, P. K. Ko, M.-C. Jeng, "BSIM: Berkeley short-channel IGFET model for MOS transistors," *IEEE Jour. of Solid-Stae Circuits*, vol. SC-22, no. 4, pp. 558-566, Aug. 1987.

[82] B. J. Sheu, W.-J. Hsu, P. K. Ko, "An MOS transistor charge model for VLSI design," *IEEE Trans. on Computer-Aided Design*, vol. CAD-7, no. 4, pp. 520-527, Apr. 1988.

[83] M. Sililotti, M. R. Emerling, and C. A. Mead, "VLSI architectures for implementation of neural networks," *Neural Networks for Computing, AIP Conf. Proc. 151*, Editor: J. S. Denker, pp. 408-413, Snowbird, UT, 1986.

[84] M. A. Sivilotti, M. A. Mahowald, and C. A. Mead, "Real-time visual computations using analog CMOS processing arrays," *Proc. of the Stanford Advanced Research in VLSI Conference*, Cambridge, MA: The MIT Press, 1987.

[85] G. D. Smith, *Numerical Solution of Partial Differential Equations: Finite Difference Methods*, Oxford University Press, pp. 60-63, 1985.

[86] B. Soucek, M. Soucek, *Neural and Massively Parallel Computers*, New York: Wiley-Interscience: 1988.

[87] B. Soucek, *Neural and Concurrent Real-Time Systems*, New York: Wiley-Interscience, 1989.

[88] S. M. Sze, *Physics of Semiconductor Devices*, pp. 500-504, 2nd Edition, New York: John Wiley & Sons, 1981.

[89] D. W. Tank and J. J. Hopfield, "Simple 'neural' optimization net-
      works: an A/D converter, signal decision circuit, and a linear pro-
      gramming circuit," *IEEE Trans. on Circuits and Systems*, vol.
      CAS-33, no. 5, pp. 533-541, May 1986.

[90] J. E. Tanner and C. A. Mead, "A correlating optical motion detec-
      tor," *Proc. of Conf. on Advanced Research in VLSI*, Dedham, MA:
      Artech House, 1984.

[91] S. T. Toborg, K. Hwang, *Exploring Neural Network and Optical
      Computing Technologies*, in Parallel Processing for Supercomputers
      and Artificial Intelligence, Editors: K. Hwang and D. Degroot,
      New York: McGraw Hill, 1989.

[92] M. S. Tomlinson Jr., D. J. Walker, M. A. Sivilotti, "A digital
      neural network architecture for VLSI," *Proc. of IEEE/INNS Inter.
      Joint Conf. on Neural Networks*, vol. II, pp. 545-550, San Diego,
      CA, June 1990.

[93] C. Tomovich, "MOSIS: A gateway to silicon," *IEEE Circuits and
      Devices Magazine*, vol. 4, no. 2, pp. 22-23, Mar. 1988.

[94] P. Treleaven, M. Pacheco, and M. Vellasco, "VLSI architectures
      for neural networks," *IEEE Micro Magazine*, vol. 9, no. 6, pp. 8-
      27, Dec. 1989.

[95] Y. P. Tsividis, *Operation and Modeling of the MOS Transistor*,
      New York: McGrow-Hill, 1987.

[96] Y. P. Tsividis, "Analog MOS integrated circuits - certain new
      ideas, trends, and obstacles," *IEEE Jour. Solid-State Circuits*, vol.
      SC-22, no. 3, pp. 351-321, June, 1987.

[97] D. E. Van den Bout and T. K. Miller, "A traveling salesman
      objective function that works," *Proc. of IEEE Inter. Conf. on
      Neural Networks*, vol. II, pp. 299-303, San Diego, CA, July 1988.

[98] D. E. Van den Bout and T. K. Miller III, "A digital architecture
      employing stochasticism for the simulation of Hopfield neural
      nets," *IEEE Trans. on Circuits and Systems*, vol. 36, no. 5, pp.

732-738, May 1989.

[99] D. E. Van den Bout, P. Franzon, J. Paulos, T. Miller III, W. Snyder, T. Nagle, W. Liu, "Scalable VLSI implementations for neural networks," *Jour. of VLSI Signal Processing*, vol. 1, no. 4, pp. 367-385, Boston, MA: Kluwer Academic, Apr. 1990.

[100] P. J. M. van Laarhoven and E. H. L. Aarts, *Simulated Annealing: Theory and Applications*, Boston, MA: Reidel, 1987.

[101] A. Vladimirescu, S. Liu, "The simulation of MOS integrated circuits using SPICE2," *Electron. Res. Lab. Memo ERL-M80/7*, University of California, Berkeley, Oct. 1980.

[102] F. M. Wahl, *Digital Image Signal Processing*, Boston, MA: Artech House, 1987.

[103] H. A. R. Wegener, "Endurance model for textured-poly floating gate memories," *IEEE Inter. Electron Devices Meeting*, pp. 480-483, Dec. 1984.

[104] B. Widrow and Bernard, "An adaptive 'Adaline' neuron using chemical 'Memisters'," *Stanford Electronics Lab.*, Technical Report Number 1553-2, Oct. 17, 1960.

[105] B. Widrow, Bernard, Hoff, and Marcian, "Adaptive switching circuits," *1960 IRE WESCON Convention Record*, Part 4, pp. 96-104, Aug. 23-26, 1960.

[106] G. V. Wilson and G. S. Pawley, "On the stability of the traveling salesman problem algorithm of Hopfield and Tank," *Biol. Cybernetics*, vol. 58, pp. 63-70, 1988.

[107] Y. S. Yee, L. M. Terman, and L. G. Heller, "A 1mV MOS comparator," *IEEE Jour. of Solid-State Circuits*, vol. SC-13, no. 3, pp. 294-298, June 1978.

[108] Y. Zhou, R. Chellappa, A. Vaid, and B. Jenkins, "Image restoration using a neural network," *IEEE Trans. Acoustics, Speech & Signal Proc.*, vol. 36, pp. 141-1151, July 1988.

[109] S. F. Zornetzer, J. L. Davis, C. Lau, Editors, *An Introduction to Neural and Electronic Networks,* New York: Academic Press, 1990.

Index